Tradesmen in Business

A COMPREHENSIVE BUSINESS GUIDE AND HANDBOOK FOR THE SKILLED TRADESMAN

Bob Rowan & Melvin Fenn

BETTERWAY PUBLICATIONS, INC.
WHITE HALL, VIRGINIA

Published by Betterway Publications, Inc.
Box 219
Crozet, VA 22932

Cover design by Tim Haley
Typography by TechType

Map of "Areas of Code Influence" reprinted with permission from *Understanding Building Code and Standards in the United States*, National Association of Home Builders.

Library of Congress Cataloging-in-Publication Data

Rowan, Bob
    Tradesmen in Business

    Bibliography: p.
    Includes index.
       1. Industrial management—Handbooks, manuals, etc.
2. Entrepreneurship—handbooks, manuals, etc. I. Fenn,
Melvin C.   . II. Title
HD31R76 1988   659   88-19417
ISBN 1-55870-103-6 (pbk.)

Printed in the United States of America
0987654321

Dedicated to the tradesman:

To those who exemplify integrity, hard work, craftmanship,

and have the self determination to build a better world.

ACKNOWLEDGEMENTS

For the technical skills, dedication and strong character provided by Bob Rowan, so pertinent to the accomplishment of this task, I am grateful.

To Doris Tigert, appreciation and gratitude are extended, for professional critique and long hours of assistance.

Thanks also to Robert Hostage, and Betterway Publishing Company, a very real necessity with class.

Melvin C. Fenn

To Melvin Fenn, for his creativity and experience as a model tradesman.

To Karen Rowan for her patience and support through the many months of working nights and weekends.

To Ron Bollinger for his encouragement and suggestions; a classy businessman and a quality individual.

Bob Rowan

CONTENTS

INTRODUCTION 9

SECTION I: BUILDING A BUSINESS

   1. Making Your Move 15
   2. Going for the Long Haul 29
   3. Technical Skills 39

SECTION II: DOLLARS AND SENSE

   4. Estimating and Bidding 51
   5. Contracts and Proposals 61
   6. Financing 69

SECTION III: WORKING WITH PEOPLE

   7. Employees 75
   8. Clients 83
   9. Subcontractors and Suppliers 91

SECTION IV: SHARPENING THE EDGE

   10. Accounting 97
   11. Legal 115
   12. Insurance 117

SECTION V: SUMMING IT ALL UP

   13. Home Environment 123
   14. Back to the Future 125
   15. Conclusion 129

SECTION VI: RESOURCE MATERIALS

   Publications 135
   Resource Addresses 153
   Bibliography 158
   Forms 159

INDEX 223

INTRODUCTION

Subcontractors and tradesmen have been coming to us for years to ask how best they could handle their business problems. Many of the same questions were asked over and over, not just by the tradesmen starting out in new business ventures but by established small contractors as well. It became increasingly apparent that many of these small business owners knew nothing about operating a business. They were all skilled craftsmen; their business problems usually had nothing to do with the quality of the work they performed. But nobody had taught them any business skills. We decided to write this book to fill that gap in their education, to help all those qualified tradesmen who want to match their practical skills with comparable business skills; to teach them how to manage their businesses.

Many tradesmen in business suffer from the misconception that once the business has started — they are doing work for which (presumably) they will be paid — their major worries are over. The truth is that almost anyone can start a business, but the ones who succeed are those who have the ability to manage it effectively. Starting a business is easy; staying in business can be a struggle. Anyone who has owned a trade-oriented business can describe in painful detail all the mistakes that were made throughout the "learning how to run the business" process. Many of those individuals are no longer in business; the mistakes made were simply too costly too overcome.

An estimated 85-90% of all new businesses fail, often in the first year. There are many reasons for this appallingly high failure rate. Often it is simply a matter of undercapitalization. Maybe the new entrepreneur has not researched his market adequately. Sometimes a business started during a boom time fails when the economy enters a downturn period. But most often, small business failures occur because the owners did not manage certain areas of their businesses adequately.

This is not a "big business" book. It has been written specifically for independent craftsmen who either are in business for themselves already or are planning to start a new venture in the near future. Thus the title of the book: *Tradesmen in Business*. We have a special interest in and compassion for TIBs. We understand their struggles and have worked through many of their frustrations and disappointments. We feel an almost personal sense of loss when one fails because he was not equipped to handle the management aspects of his business.

We want to increase the TIB's knowledge of tested business practices and principles, business management methods that worked successfully for other small contractors. We want to help skilled craftsmen become skilled businessmen, help them develop skills as business managers that match their tradesmen skills. In this book we carefully explain all the elements of a successful business; start-up, controlling spending, submitting bids, preparing contracts, hiring subcontractors, working with clients, hiring and managing employees, keeping records, and taking care of taxes, insurance, and many other business details that become problems if they are not properly managed.

Developing management skills takes time and effort. Our intent with this book is to help TIBs build a sound business; not a fast-growing, quick profits enterprise but one that will provide a good and growing income for years. We will emphasize building a business slowly, but steadily; a business that will grow because its ingredients for success are the application of craft and management skills, honesty, integrity, and hard work.

SECTION I
BUILDING A BUSINESS

1. MAKING YOUR MOVE

GETTING STARTED

Planning and preparation. We all understand these two words but seldom show any respect for them. Very few tradesmen in business started out after a period of wise planning and careful preparation. Most of us saw what we perceived as a great business opportunity and stepped out on our own. Once that first opportunity had been met, we looked around and said "now what?" This is not the time to panic, or to quit, but it is the time to develop a better understanding of the business we are in and to begin to apply sound principles and practices to the way we conduct that business in the future. From this moment on, every business step should be taken only after thoughtful planning and preparation.

Start on a Part-time Basis

If you are currently employed as a craftsman, don't decide one day to quit your job and start your own business (no matter how skilled a craftsman you are or how admired your work is). It takes time to build and develop a business. More often than not it takes a long time for a business to become profitable enough to support you and your family. If you want to become an independent businessman, start on a part-time basis. Sacrifice some of your evening and weekend free time by taking small jobs. You can begin to make a name for yourself this way, without sacrificing your regular paycheck.

No company, your employer's or any other, can handle all the work that is available. Often the large companies are not interested in bidding the smaller jobs, so there should be opportunities for you. You will want to proceed with caution and integrity. No employer is interested in having its employees compete with them for work or in having its employees expend time and energy on sideline jobs. They have every right to expect you to be a loyal and faithful employee.

If you value the relationship you have with your employer (and you should) you will want to be open with him about your eventual desire to venture out on your own. Let him know you value his advice. Building constructive business relationships can be extremely important to you over the years. You will want you current employer to be an ally, not an enemy; one who refers you to jobs he is not prepared to handle, not someone who will give you negative references. Burning bridges is, almost always, a bad idea. At this very early stage in your business development, it could be devastating.

There is always the possibility that the employer will terminate or otherwise penalize any employee who takes on side jobs, whether it is just for the additional income or training for future business ownership. In such a situation, it would be difficult for you to acquaint your employer with your plans to be independent. You may have to make your sideline activities a private matter, but be sure that whatever you do on your own does not affect negatively the work you do for your employer.

Use both professional courtesy and common sense, and do not compete with your employer in your sideline business activities. A conflict of interest could lead to your dismissal and a broken relationship that might never be mended. Respect your present employer and your relationship with him.

If your employer has endorsed your going into business for yourself, be sure that contacting his clients is always done with his awareness beforehand. Any job you obtain using privileged information might not only jeopardize your present employment, it could have a negative effect on any long-term relationship you might have with the client. Establish your reputation for honesty and integrity from the outset. People admire these qualities in the people with whom they do business. As your reputation becomes known, clients will seek you out because of those qualities.

One way to eliminate any possibility of bad feelings between you and your employer is to offer him a fee or commission on any jobs you obtain through his contacts. If he suggests this, or agrees to it, there are a couple of things to watch out for. First, be sure your bids include his commission or he will be the only one profiting from the work. Second, even though your employer is being compensated, he still may want to limit your outside activities. Don't overstep your boundaries. Establish with your employer what your outside limits are and make sure you stay within them.

Start with small side jobs. Get a moderate flow going for a couple of months to see how you can handle working days, nights, and weekends. If you are still interested in pursuing your dream, and the work is available to you, consider working for someone else on a part-time basis and work for yourself full-time ... still providing you and your family with a secondary income while you work to establish the business.

Becoming a Tradesman in Business is not easy. It takes determination and hard work. There are no get rich quick schemes. You can make it as an independent businessman. You probably will not get rich, but you will be able to make a good living—and you will have all the emotional and psychological rewards that there are for the person who succeeds in business for himself.

Company Name

What's in a name? Probably more than you realize. The name you give your company should accomplish two things: it should describe what the business does and/or who owns it; and it should look and sound professional. If drywall tradesman Charles Q. Bond wanted to form his own company, he might call it Bond Drywall Contractors.

Name recognition is important. If Mr. Bond had named his company Alpine Drywall it might have sounded exciting and distinctive, but it would not have identified the man behind the company. After all, Charlie Bond is trying to build on the reputation he has established for himself as a man who does quality work. A TIB sells himself and his abilities; as a result, clients associate his name with his work. Using your name as part of your business name not only indicates your confidence in yourself and your commitment to your company, it tells the client who is in charge.

As your business grows, you may want to consider expansion into other areas. Certain company names might limit your ability to grow and diversify. For example, Smith Sheetrock Hangers could be limiting if Smith wants to branch out into other fields. Try to give your company a name that leaves room for growth. Changing the business name later to accommodate changes in the business itself may prove to be both difficult and costly. Try to plan ahead and get it right the first time.

You have your name. What now?

It's time to handle the technical details, all the paperwork involved. When you start a business, you must complete a government form called an "Assumed Name Certificate." These usually are available at the county clerk's office. When you have completed and submitted this form, a name search will be made to see if another business, already established, has the name you have selected. If the name you have chosen is available, it will be recorded and filed by the county clerk as your legal business name. Usually a small fee is charged for this service.

Your next step is a call to the local Internal Revenue Service office, where you apply for a Federal Identification Number. IRS will send you a Form SS-4, with an instruction sheet. (The form number may change from time to time, but the IRS will send you the form you need.) Complete the form and return it to the IRS. In about six weeks, you will receive your federal identification number. You also will be put on a mailing list to ensure you receive the necessary forms and instruction booklets for quarterly payments, taxes, etc. The ID number should be used on all the forms you fill out for the IRS. Refer to the chapter on taxes for more information, as needed.

LEGAL FORMS OF BUSINESS

There are three basic forms: sole proprietorship, partnership, and corporation. Each offers advantages and disadvantages. It will be up to you (with the guidance of your lawyer and accountant) to determine which is best for your particular situation.

Sole Proprietorship

Most TIBs start out as sole proprietorships, principally because that is the simplest kind to start, operate, and — if necessary — terminate. A sole proprietorship is a business owned entirely by one person. No legal agreements need be written to start the business. The TIB will be wholly responsible for every aspect of the company's operations; all the

Legal Forms of Business

	Pluses	Minuses
Sole Proprietorship	1. Controlled by owner 2. All profits to owner 3. Little regulation 4. Easy to start 5. Earnings personally taxed	1. Liability unlimited 2. Limited resources 3. No continuity at retirement or death
General Partnership	1. Joint ownershop and responsibility 2. Access to more money and skills 3. Earnings personally taxed 4. Limited regulation and easy to start	1. Conflict of authority 2. Liability unlimited 3. Profits divided 4. No continuity at retirement or death
Limited Partnership	1. General partner(s) run the business 2. Limited (silent) partners have no liability beyond invested money 3. Profits divided as per partnership agreement 4. Earnings personally taxed	1. Limited partners have no say in business 2. General partners have unlimited liability 3. More regulations than general partnership
Corporation	1. Limited liability 2. Ownership interest is transferrable 3. Legal entity and continuous life 4. Status in raising funds	1. Regulated by states 2. Costly to form 3. Limited to chartered activities 4. Corporate income tax plus tax on personal salary and/or dividends
Subchapter S Corporations	1. Receives all advantages of a corporation 2. Electing corporation taxed as sole proprietorship	1. Highly regulated both by state and IRS 2. Restricted to certain kinds of business and limited number of stockholders

The above information is reprinted from the U.S. Department of Labor booklet, "*More Than a Dream: Raising the Money*."

operating decisions, all the profits earned, all the debts incurred. As the sole proprietor, the TIB is his own boss. There are no partners with whom to share decisions, no board of directors to whom to report. Decisions can be made quickly, without interference. And the cost of starting up in this mode is minimal. The legal fees that accompany setting up a partnership or corporation are not involved with the formation of a sole proprietorship.

However ... there are disadvantages to this form of business and those disadvantages may encourage the TIB, at some point, to change the business to a corporation. When a TIB is functioning as a sole proprietor, he may find raising capital, from lending institutions or investors, to be very difficult. If the TIB has his own, adequate start-up capital, the inability to raise funds would not be a problem. For those who are starting with limited funds, it could be.

An even more critical disadvantage to the sole proprietorship is the fact that the TIB is fully liable for all business debts and for the negative consequences, if any, of his actions. That means that his personal assets are not protected from legal and/or financial claims. If the TIB, operating as a sole proprietor, is sued (for whatever reason) or his business fails, his personal assets can be seized to satisfy any claims and/or debts.

Notwithstanding these concerns, the sole proprietorship is normally the best way to start a new business. As the business grows, and the risk of having personal assets unprotected increases, the business can be converted to a partnership or corporation.

Partnerships

The Uniform Partnership Act definition of a partnership (accepted by more than three-quarters of the states) is: "An association of two or more persons to carry on as co-owners of a business for profit."

There are two types of partner: general and limited. A general partner is one who takes part in the day to day management of the business and has unlimited liability (financial obligation).

Limited partners cannot perform any of the managerial functions of the business. If they do, they will be recognized by the courts as general partners (thereby increasing their liability for the business). The most common reason for the formation of a limited partnership is to provide the managing partner with increased operating capital. Individuals who want to invest money or other assets in a new company but who do not want to assume any management responsibilities, may find a limited partner role attractive. They risk their investment but have no additional liability.

Although the law does not require that persons involved in a partnership have a contract or written agreement, it is very important that there be one. Misunderstandings, hurt feelings, even relationship-dissolving arguments may arise if some form of agreement is not used. Such a partnership agreement should include the following:

1. The name of the firm and of each partner.
2. The type of business and its location.
3. Duties and authority of each partner.
4. The duration of the partnership agreement.
5. The amount of money and/or other assets invested by each partner.
6. The way in which net profits or losses will be divided.
7. The way in which each partner will be compensated.
8. Limitations on the extent to which funds can be withdrawn by either partner.
9. The accounting procedures that will be followed.
10. The procedure for admitting new partners.
11. The procedure for dissolving the partnership.
12. The signature of each partner to the agreement.

All property that is invested in the partnership by one partner becomes the property of all partners jointly. There should, of course, be a record in the books indicating the amount that each partner has contributed. If and when the partnership is dissolved, each partner's claim against the assets of the business will be governed by his percentage of the overall initial investment.

As noted above, the way that profits and losses will be distributed should be noted in the partnership agreement. If each partner has contributed equally in capital and/or services, there probably would be an equal division. However, if one partner has contributed services and/or capital that exceeds in value the contributions of other partners, the division of profits and losses should reflect those investment percentages. If there is no agreement as to the distribution of income and losses, all partners will share equally.

By its very nature, a partnership has a limited life. If a partner ceases to be a member of the business for any reason — bankruptcy, death, withdrawal for other reasons — the partnership is dissolved. Also, if a new partner is admitted to the firm, a new partnership must be formed, with a new partnership agreement.

All partners must agree before a new partner is added. While an existing partner can sell all or a part of his share of the business to another person, and that new person has the right to share in the profits, losses, and assets of the business, no new person can automatically become a "partner" unless or until he is admitted to the firm by all other partners.

It is easy to see the usefulness of a partner in your business. He may be able to provide valuable knowledge and experience or may have the financing resources to ensure a stable future for your company. A partner may have just the skills and qualifications you are lacking. On the surface, going into business with such an individual seems to make a lot of sense.

However, like marriages, not all partnerships are made in heaven. You may end up with more — or less — than you bargained for. A partnership is a commitment, between or among two or more individuals. Most individuals have different motivations, abilities, desires, and goals. Sometimes these differences result in conflict. But not every partnership has such problems and not all partnerships are destined to fail because of such differences. If you are now a sole proprietor and are considering changing your business to a partnership, weigh the advantages and disadvantages of your current situation with the advantages and disadvantages of sharing investment, responsibilities, and control, and make the decision that best satisfies your needs and interests. And know

as much as you can about your prospective partner; his abilities, his personality, his character, and his financial situation.

In the beginning, in most partnership relationships, things are great and everything is running smoothly. But one day you have to buy a piece of equipment ... or fire an employee ... or borrow money ... or make a business-affecting decision. Whatever it was, your partner may feel that your decision was wrong. You cannot understand his reasoning; in fact, his point of view makes no sense at all to you. The next week a similar situation develops. Then another. Then that kind of dispute seems to become a daily occurrence. Pretty soon you can't stand being around each other. This kind of situation develops more often than not. Oddly enough, it develops most frequently when the business is thriving and successful.

Understanding the Division of Responsibilities

In any business, one person must be in charge. Everyone in the organization must be subject to his authority, or the normal day to day decision making processes will function properly. If you have a partnership, grant one person the controlling interest so he can make decisions without interference. Be sure that each partner understands his partnership responsibilities. Each should prepare a letter of intent stating his goals, assumptions, understanding of the risks, business growth projections, understanding of the responsibilities of the other partner, etc.

As the business grows, it also changes. The direction may change, financial conditions may change, and the interests and responsibilities of the partners may change. If you enter into a partnership without (1) a full understanding of the potential problems involved and (2) clearly understood plans that can accommodate a variety of changes, you probably are ordaining your own future failure. You should have pre-established plans and procedures that will help each partner move easily into changing roles if future developments require such changes.

Thoughtful planning and preparation will always give

you a better chance for success. Partnerships are not wrong for everybody; they can succeed and endure, but understand fully what you are getting into — and with whom.

Corporation

A corporation is the most complicated form a business can take. While it offers special advantages, such as protection for one's personal assets in the event of a law suit, it involves a lot paper work, legal fees, and accounting fees.

By definition, a corporation is "an association of individuals, created by law and existing as an entity with powers and liabilities independent of its members." (Random House College Dictionary). In simple language, a corporation is completely separate in its business and financial dealings from the non-business activities of the person or persons who formed it. The corporation has the same rights as individuals, such as the right to own and sell property, the right to manage its affairs, and the right to sue. Like an individual, a corporation also can be sued.

Forming a corporation is both more difficult and more costly than forming a sole proprietorship or a partnership. While there are how-to books on forming a corporation, you probably will need qualified legal counsel to ensure that you comply with all state laws and procedures.

A corporation must have a board of directors; a minimum of two or three persons, depending on the requirements of the state in which you do business. The directors can be anyone the TIB chooses, including himself. The board of directors must have meetings at least once a year at which written minutes are taken to document what was discussed. The directors have the responsibility to give direction to the business; to evaluate current operating practices and to help set long term business goals. They should review the financial condition of the business and provide suggestions as to how the business is to be structured for profitable growth.

There are various state and federal government reports that corporations must submit. Federal income tax returns will have to be filed. Also many states (and counties) impose an annual franchise tax or business license fee on all corporations doing business within their borders.

Subchapter S Corporation

This is a corporate structure for new or limited income businesses. Subchapter S corporations are corporations that elect not to be taxed as corporations. Corporate taxes can be substantial; Subchapter S brings the TIB some tax relief. Its income is taxed as though it were the TIB's income. But the Subchapter S corporation is similar to a partnership only in that respect. In other ways it is legally a corporation. Its owners have full corporate protection against liability for the debts of the business.

If your business involves some possibility of personal injury (as does the construction business), and/or you want to limit the exposure of your personal assets, or you want to avoid the risks and complexities of a partnership, a Subchapter S corporation may be right for you. Consult your attorney and your accountant before making a final decision.

LICENSES AND CERTIFICATION

Summary of State Regulations and Laws Affecting Contractors is a publication available from the American Insurance Association (85 John Street, New York, NY 10038). This booklet includes information pertaining to state legislation and to regulations and licensing requirements for all the states. County and municipal requirements are not covered in this AIA publication, but that information is easily obtained from your local building permit office.

About half the states require some form of licensing or registration for general contractors and/or sub-contractors. Some states will require the contractor to take a written examination covering a broad range of subjects; license law, business management, ability to read plans and specifications, cost estimating, ethical standards, and other matters that relate to the specific trade for which a license is being sought. Even though there may not be any state law requiring licensing, most cities do require a license for specific trades.

Some states also require that surety or cash bonds be posted. Some also require proof of liability and property damage insurance.

It is very important that you investigate and find all the laws that apply to your trade in the state and municipalities where you intend to bid work. You also should make yourself aware of any laws or regulations that grant preference in the award of public contracts to bidders who reside in the state in which you plan to bid work. You will need to have all the relevant information on any local sales taxes, income taxes, and codes and ordinances of all the municipalities and counties in which you will be performing work.

ORGANIZATIONAL STRUCTURE

The importance of establishing a system of organization in every phase of the TIB's business cannot be overemphasized. However, it is unusual to find a TIB who has recognized the business as being a "system" ... a system that can be organized and built upon so the business, as it grows, becomes more than just "Joe Smith, Plasterer." It becomes a functioning system. It becomes an organization in which the TIB's principles, goals, ideals, and standard procedures are not just known familiarly as the boss's thoughts and ideas, they are established as the standards of the business for every employee.

This may seem a difficult and complicated chore. It is not. It is one of the most critical aspects of business management. Unfortunately, it also is one of the least understood. Remember, the aim of a TIB is to become a competent business manager. Developing a business system is part of that process, an important step that should be taken very soon after the business is established.

It is a matter of fact, of course, that most TIBs start out doing most, if not all, of the work themselves. From preparing the bids to doing the work, from keeping the records to hauling away the trash, the TIB is the business. Without him, there is no business. For many TIBs, this will always be the case. But many others who want to grow have a hard time getting

over the hump, the transition from being a one-man operation to managing a growing, functioning enterprise with a credible growth plan. A TIB's company is not apt to be an instant money-making success, but it can and will grow if the right business tools and judgments are applied. If the TIB is a skilled tradesman and if, as he grows, he employs other skilled tradesmen, increased business revenues are almost guaranteed. With a system of business management, quality control, client satisfaction, and profitability also will be assured.

Inevitably, most new businesses are organized around personalities and immediate needs rather than job functions and standardized procedures. In other words, around people and coping with emergencies rather than doing the various "routine" jobs that must be performed if the business is to run smoothly. For example, as the business grows, there suddenly is more work than the TIB can handle. He needs help, so he hires a couple of helpers. He isn't exactly sure what needs to be done, or by whom, but new employees begin working at the job as laborers and general helpers and are called upon to perform other specific functions as needs arise. The new employees, usually eager to please, help with completing the work, ordering necessary materials, picking up those materials, meeting clients, keeping records, hauling away the trash, etc.

For a while, things seem to proceed pretty smoothly. If the TIB doesn't have the time to carry out a particular task, one of the other guys does it. But then the TIB begins to notice that some of the small details — items he would handle routinely — are not being managed completely: orders are not being carried out as given; material is not being ordered economically; even the trash is not being taken away correctly. Most important, clients are complaining about mistakes on the job.

What happened to this TIB is common. Because he did not have in place an organized business system, he lost control of the business. A working business system has a positive and dramatic impact on the life of the TIB. It allows him to control his business ... not vice versa.

How to Set Up the System

The first step in the process of establishing a business system begins with the organizational structure. A very simple, but critical, action is part of this first step — and part of everything else the TIB does to manage his business. He writes everything down. Every part of the business system is put on paper. Verbal orders and guidelines can supplement the written words, but those written words must be the basis of the system.

First, every major function of the business should be listed: supervising the work; purchasing materials; picking up materials; equipment maintenance; job clean-up; warranty work; and the various operations of the trade. Next to each of these job functions, the various responsibilities that are part of each job should be listed. The TIB should be as specific and detailed as possible here, as if he were explaining to a new employee how to carry out each task. Some jobs will require more explanation than others, but it is important to cover each one thoroughly. Over time, as he performs each task himself, the TIB will see how new insight can be added to the job descriptions. This ongoing process of revision should make the list of responsibilities as clear and comprehensive as possible.

The next step is to identify each job description with either an employee's name or the name of the TIB. The individual so designated becomes fully responsible for that job. He may not be the only one doing it, but he is the one held responsible for making sure it gets done correctly. Of course, every employee will report to the TIB and be subject to his supervisory control.

In every start-up or young business, the TIB himself typically is the individual responsible for every job function. However, as the business grows, needs arise, and new employees are hired, the job descriptions established by the TIB when he was the only employee will help each new worker understand his responsibilities, job boundaries, and level of authority. When workers understand their levels of responsibility and accountability, they are able to function efficiently and without making the TIB unduly concerned about who is going to do what.

Operational Procedures

Establishing a list of operational procedures is the next step in the development of a business system. In other words, you need a set of instructions for each employee that explains what he should do, how he should act, in every situation. Do your employees know what to say when a client shows up on the job unexpectedly? What to do when a worker is injured? On the negative side, do they know what actions will get them fired automatically?

Let's say, for example, that a client visits one of your jobs. The expression on his face may not show any emotion, but it's likely his subconscious mind is making an inventory of everything he sees. Is the job site clean? How are the employees dressed? How do they act? What do they say? Each one of these factors could influence the client's decision to rehire the TIB for a future job. If your workers know how to act, what to say, and are reasonably dressed (appropriate for the work they are doing), they will be effective in having a positive influence on the client. If they handle themselves in a businesslike way, the business not only will look efficient, it will be efficient.

Operations Manual

To accomplish this level of efficiency consistently you will need an Operations Manual; a set of specific rules, guidelines, and instructions covering such topics as clients at the jobsite, dress requirements, work schedules, unacceptable behavior, operating equipment, tools, job safety, increases in compensation, employee benefits, etc. The manual should be as comprehensive as possible, and expanded as new situations develop because of the company's expansion and diversification. Most companies refer to this kind of package as their operations manual. Each employee should be given a copy as soon as he is hired and advised to familiarize himself with its contents as quickly as possible. If you have hired an employee with the appropriate job skills, he should be able to perform effectively almost immediately, according to your rules and policies.

One of the keys to having good employees — satisfied workers doing their work well and representing your company well — is your understanding of human nature; a trait you must develop if you don't possess it intuitively. Your employees want to work according to clearly defined operational procedures; a business structure within which they can work toward clearly established objectives, with work and behavior standards that are sensible and reasonable. They want to know who is in charge of what and who reports to whom. This approach may seem restrictive to you, but actually it has just the opposite effect. It gives each employee the freedom and challenge to be the best he can be, playing by the rules of the game.

Picking Up Speed

It is important for a TIB to bid enough work to keep him and his employees busy. If the TIB himself is actively involved in doing the work, and has only one or two employees, determining the ideal work load should not be difficult. However, if only four "average" jobs are needed each month to create enough work and a good income, the TIB should not assume that only four jobs need to be bid. The TIB will get a feeling, over time, as to how many jobs he will get as a percentage of the number of jobs he bids. Typically, one out of ten — a 10% closing rate, is considered pretty good. A TIB who is determined to be successful (particularly in the start-up years) must pursue as much work as possible.

Anyone who has ever been out of work knows that looking for work is, by itself, hard work. Many TIBs have failed to succeed because they did not invest the time and effort necessary in looking for and bidding new business.

A heavy workload is the most desirable situation. It generates enthusiasm and motivates everyone — the TIB and his workers — to work to the highest and most productive levels possible. There will be times, however, when the size and scope of the workload will look overwhelming; when the "team" is more or less intimidated by the amount of work to be done. At times like these, the TIB must assert his leadership, inspiring a "can do" attitude through his own positive approach and confidence. Truism and clichés can be boring ... but they also tend to be accurate. Experience IS the best teacher. As the TIB learns his business, he

will be able to determine how much and what kind of work he needs to provide both adequate work flow and income.

It Won't Be Easy

Anyone who owns his own business makes a lot of sacrifices, the greatest one being the long hours worked. After putting in a long, hard day, the typical TIB can expect to come home to a nightly routine that will include phone calls, preparing estimates, paying bills, and a variety of other tasks that have to be tackled to keep the business moving. Twelve to fourteen hour work days will not be unusual, particularly during the "getting established" period. But the strong will survive and those excessive time demands will ease over time. No TIB, however, should ever anticipate a routine life of nine to five work days.

No TIB should ever get overcommitted to the point that he can't provide his clients with quality work performed on time. If quality is lost, everything is lost. Make sure that finishing a job in a rush does not jeopardize the quality of the work. Always be straightforward with the client. Give him a reasonable timetable for the completion of his work. If slippage from the original schedule is anticipated, it is best to "face the music" as soon as that is known. Tell the client immediately, rather than having to apologize and explain when the original completion date arrives and the work is not close to being finished. Without client satisfaction, you can't possibly build a successful business. Dealing with each client straightforwardly and honestly probably is as important as the quality of the work your company performs.

Consistently providing your employees with regular hours is a must. And while there will be times when you will have to ask them for an extra effort — maybe working long hours several days in a row to finish a particular job — the TIB will not be able to impose that requirement on a regular basis. If you try to, and your employees start to wear down from stress and fatigue, you will notice an accompanying dropoff in motivation, dedication, and quality of workmanship. Sometimes it is a good idea to ask your employees if they want to work overtime. If they are willing, and the time/stress demands are not too great, sticking with your "regulars" usually works better than hiring additional help you may have to release when the work load levels off.

Owning and managing your own business can be rewarding for other than financial reasons. The feeling of accomplishment is very satisfying. Even the sacrifices you make will enrich your life if you manage your time effectively and learn to eliminate the activities that are non-productive. Hard work and dedication in the beginning stages of the business should give the TIB the kind of good performance record that makes work easier to obtain in the future. The business may not get easier, but it should get better.

TIME MANAGEMENT

TIBs fill their days (and some nights) taking care of such responsibilities as scheduling workers, ordering materials, picking up materials, checking jobs, working jobs, meeting clients, bidding jobs, completing paper work, and making phone calls. Many of these tasks involve traveling from one place to another. They provide many opportunities to waste time and effort through poor planning and scheduling. Clearly, controlling his time is critical for the TIB.

You have noticed that some people work ten-twelve hours a day without accomplishing a whole lot. Others seem to accomplish twice as much in half the time. Why? They manage their time effectively. By definition, time management is the "act of controlling events." Unfortunately, most people (including many managers) go through their days letting events control them. How most business responsibilities are handled is described by the following three personalities:

The Fire Fighter. This person spends his entire day dealing with small emergencies. As a problem arises, he runs to take care of it. As soon as one problem is solved, another comes along to demand his attention. This person's days are filled with activity, but the really important things never seem to get done.

The Freedom Fighter. This guy just goes with the flow. He takes care of whatever comes along. Noth-

ing is a big deal. He responds only when the situation demands his attention. He has no plan and virtually no control over his business.

The Time Manager. The time manager understands that "daily urgencies" can distract him from accomplishing those things that are vital to the productivity and success of his business. He sets goals and plans ahead. He controls the business; he doesn't let it control him.

The first step in managing your time is finding out how you are spending it. Commit a two week period to recording exactly how you spend each day. Be as detailed as you can in keeping this time record. Measure your time in minutes, the way lawyers do (for billing purposes!). At the end of the two weeks, analyze how your time has been spent. Once you are aware of that, you should be able to eliminate the unnecessary activities that have been overloading and cluttering your day.

Probably the most important ingredient in any time management program is spending the first 15-30 minutes of each workday planning and scheduling that day. A rule of thumb is that each hour of planning saves five hours of time needed to complete the tasks at hand.

Time management consultants always advise you to "make lists", and organized people tend to be list makers. Without a list, you may feel there is so much to do there is no way you could handle all of it. With a list it somehow seems possible. So use that morning planning time to make a list of everything that must be done that day; individual, achievable daily goals. Then prioritize all the items on the list, rating each item on the basis of its importance. If you don't set priorities every day, you probably will find you go through each day doing most of those items anyway ... except for one or two things. Those few omitted tasks — jobs that did not get done — may have been that day's most vital events.

When you are setting priorities, you'll find that the familiar 80/20 rule will apply. Marketing professionals know that twenty percent of their customers will give them eighty percent of their business. In the context of time management, the 80/20 rule means that twenty percent of what you do will generate eighty percent of the results your company achieves. The key, of course, is to identify those twenty percent activities and make them your top priorities.

One way to organize your lists is to identify each item as having one of the following priorities:

1. Vital. These are your "A" priorities. No matter what else happens that day, you take care of the "A" items. If your daily list includes more than one "A" item, identify them in order of importance or schedule sequence as "A1," "A2," etc.
2. Important. These are the "B" events. They are important and they must be taken care of, but they have less value or impact than "A" items. These, too, should be identified in order of importance as "B1," "B2," etc.
3. Limited Value. These "C" items should be handled sooner or later, but they have limited value. "C" tasks should be addressed only after the "A" and "B" items have been resolved.
4. Waste. This category includes those things that don't really have to be done. If everything else has been handled, and time permits, these items can be considered.

Time Management Planning System
In addition to preparing daily "Things To Do" lists, the TIB also has appointments to remember, phone numbers to keep track of, and "future planning" notes to make. The proper tools simplify the managing of these aspects of the daily schedule. Fortunately, there are several excellent planning systems available, either by mail or from your local stationery/office supply retailer. They are:

Day Timer by Day Timers, Inc.
(800) 556-5430
(800) 538-0861 (PA residents)
Day Runner by Harper House, Inc.
(800) 232-9786

Using these systems requires determination and self-discipline, but the rewards make the effort well worthwhile.

The planners come in a variety of sizes, from coat pocket size to a large notebook. Contact the companies for free brochures or check your local retailer. The planner offers a simple, practical method by which things to be remembered — activities, projects, people, places, expenses — are recorded for quick recall at a later date. It is a work planner, appointment book, time log, telephone and address directory, and expense record. It enables planning by the day, week, month, or year.

Some Useful Time-Saving Tips

1. Organize your home and work space. You waste time every time you look for something that is not where it should be.
2. Handle papers once only. Instead of picking up papers several times, force yourself to make a decision the first time and act on it. If you cannot do that, file the papers or throw them away.
3. Take work with you to finish so you can take advantage of time that would otherwise be wasted waiting for someone or something.
4. Establish a regular work schedule. Begin work at the same time each day. Maintaining a regular schedule will give you a greater ability to organize and schedule your day.
5. Learn to do more than one thing at a time and develop accelerated production techniques on all your routine jobs.
6. Be on time. Respecting the time of others will impress on them the importance of your time.
7. Shorten phone calls and visits. Those wasted minutes of unproductive conversation can impact your productivity.
8. Set a time limit for meetings and appointments. When a meeting starts, tell the other party how much time you have so that he will be prepared, with you, to finish within the appointed time.

Time management is the control of time and events. The control comes from accomplishing priorities, not in putting out fires. It is identifying wasted time and developing methods and habits that turn those wasted hours into productive hours. Becoming a business manager involves more than just having your own business. It means that you have learned how to manage every aspect of your business; people, finances, your product and service performance ... and time.

WHERE TO FIND WORK

There are several avenues to explore when searching for prospective jobs or clients. One of the best ways to find work or increase your clientele is to drive to the different job sites in your area and walk onto the jobs. Ask for the name of the owner, builder, or general contractor and make personal contact with him. Travel from job to job until you get results, and repeat the process as often as needed. Follow up with a phone call or two. Ask the potential client if you can call back and when is the best time to call. Following up on phone calls will reflect, in the client's eyes, how well you will follow through on the job.

Job Site Visits

Visiting the potential client's jobs just before your particular trade is about to start can many times produce a job or two. Sometimes the contractor or tradesman who normally does work for the client may be tied up or has been giving the client problems. Your presence may be just what's needed to land the job.

Previous Clients

After frequent contacts with the client, he will see how concerned you are about getting his business. It is important for him to have the confidence of knowing that you want his work and that you will give it your very careful and personal attention. Many times TIBs, after working for a client for a long period of time, will become lazy and careless assuming that getting the work is a sure thing. Unfortunately, the client doesn't feel the same way and might be in the process of looking for a new contractor when you show up on the job.

Referrals

There are a variety of ways to find out about available work. For instance, contacting past employees and associates can many times produce valuable leads and referrals. Don't let any lead or referral, no matter how insignificant it may seem, go without immediate attention. When given information, always follow it up

with at least a phone call. If your associates find out that you do not follow up on leads, they could permanently cut off the flow of information. Every referral that a businessman gives will reflect on his own business, and he will only give a good reference if he knows that you will handle the information with the proper care and attention.

Material Suppliers

Sometimes material suppliers can provide valuable leads by clueing you in on upcoming work. They will know when large deliveries are scheduled to go out to certain jobs, and customers will frequently ask the suppliers for names of qualified tradesmen whom they would recommend. This can create a valuable contact as you develop a closer relationship with the supplier, and as you prove yourself to be a faithful customer who pays his bills on time.

Banks

On occasion, your banker can provide a strong lead. This probably won't be a regular source of information because of the large number of customers he has to serve, but try to get what you can from time to time. If he is sure of your abilities, he will probably help as much as he can.

Advertising

Advertisements in newspapers and circulars can provide some response, but don't spend much money unless it has proven itself to you to bring results. Contrary to popular belief, advertising is not always effective. However, sometimes having your name and a small ad in the yellow pages of your local telephone book could bring in some business from homeowners or small contractors. This ad can be very expensive, so be conservative. And make sure that it will definitely help your business before you spend any money.

Finding work will be tough for any new TIB, and in the early days a client may show little or no loyalty. However, after a period of time and after you have proven yourself, getting work will become much easier.

Dodge Reports

McGraw-Hill Informations Systems Company, through its F. W. Dodge Division, publishes the

"Dodge Reports." Dodge sends out reports to its subscribers on construction projects out for bidding in your area. The reports are sent out every two to three days giving updates and vital statistics for each job. There are a variety of ways to receive Dodge Reports. They can be ordered by county or by groups of counties, covering the entire continental United States. Subscribers can order the reports based on over sixty types of building categories from small remodel to large high rise. There are over 120 Dodge Division offices around the country with an office in most major cities.

Unfortunately, the TIB must pay an annual fee for a subscription. This fee varies depending upon the number of counties reported and the classification of jobs. It is very expensive ($800 — $2500 per year) and prohibitive for most TIBs. Without a large flow of work the expense may not be worth it. The local Dodge office can provide more information.

GOVERNMENT AGENCIES

There are a variety of informational resources available to the TIB. The federal government, for example, has established a group of agencies designed to provide valuable information to the small business owner.

There are any number of government offices with which you can get in contact on almost any subject. If you need an expert on any topic, contact the Library of Congress National Referral Center. They have lists of thousands of specialists in government, on university staffs, and working for nonprofit organizations. The Federal Information Center, which has offices in about a hundred cities, will have similar lists of experts within the government only.

Census data can be obtained by contacting the Data Users' Service of the Bureau of the Census. They can give you information such as types of housing, family incomes, types of businesses, ages of residents in certain areas, and much more. Census information can be very useful in deciding if your trade is needed in certain areas. The Small Business Administration has a pamphlet entitled "Using Census Data in Small Plant Marketing" that may help you take advantage of the data available.

Small Business Administration

The Small Business Administration is a federal agency specifically designed to help and advise the small business person or someone considering becoming a small business owner. There are many factors that determine who is eligible for assistance from the SBA. There is no specific number of employees or size of business used as a measuring stick; it varies according to the type of business: manufacturers, retail and service businesses, steel mills, construction companies, etc. Any construction company grossing under $9.5 million annually will generally qualify.

The SBA not only grants loans for the small business unable to obtain a loan elsewhere, but will also help in selling your service or product to some of the many government agencies dealing in those services. They can also assist you in getting subcontracts for defense and space programs. The SBA has a "disaster lending" program aimed at those small businesses which have been damaged by a natural disaster through no fault of their own. These loans range from $250.00 to $250,000.00.

The SBA has an "Answer Desk" where experts from their Office of Advocacy staff the phones from 8:30 A.M. to 6:00 P.M. Monday through Friday. They welcome questions from business owners concerned with governmental regulations and training and counseling services available, as well as questions about the availability of SBA loans. The Answer Desk telephone number is 1-800-368-5855.

You can call your nearest SBA office and ask them to send you a list of some 250 free booklets and a separate list of publications sold for $1.00 to $3.00. It might be a wise idea to stop by the SBA office and look through the pamphlets that you're interested in and maybe talk with some of the people in the office, even if you don't have a particular problem at the time. It's always good practice to make friends and have a general knowledge of resources available before the need arises.

SCORE

Another source of information that can be beneficial to you and your business is an organization called SCORE. This is a business counseling service working in conjunction with the Small Business Administration. SCORE stands for Service Corps of Retired Executives. These are retired volunteers located in 592 offices in every state who offer their expertise in all kinds of services: accounting, law, business management, real estate, insurance, inventory control, plant management, site location, buying and selling, personnel, and many other problems you might encounter.

More than 300,000 small business people each year come to SCORE for business counseling. This service is provided free of charge, so do not hesitate to seek their help if you feel it will benefit you. They can help entrepreneurs on three levels:

1. When a person wants to go into business.
2. When a person with an existing business wants to expand or merely wants an objective outsider's views on expansion ideas.
3. When the business person is in real trouble.

To contact the SCORE office nearest you, look in your telephone directory in the white pages or in the government blue section under Small Business Administration. The national SCORE telephone number is tollfree: 1-800-368-5855, or write: Director of Communications, SCORE, Suite 410, 1129 20th Street, N.W., Washington, DC 20036.

2. GOING FOR THE LONG HAUL

KEEPING COSTS DOWN

As elementary as it may seem, the root cause for many failed businesses is overspending. Overspending is caused by a lack of control over and understanding of basic principles that can mean long term success. It's easy to operate with short term goals in mind. But for a lasting, successful future the key is to eliminate the tendency to overspend. When a TIB has several jobs and is "making money," he sometimes feels compelled to expand by buying vehicles, opening an office, hiring a secretary, and/or buying new equipment. Sure, these things are great to have, but every major expenditure must be made based on available working capital, a profitable workload, and a clear understanding that today's economy is constantly changing. Most TIBs clearly do not count the costs, and they buy before they can afford to or when the economy is in a downturn.

Bigger is not better, it's just bigger. Some TIBs buy out of a need to impress clients and other tradesmen in their field, and some buy out of a desire to be more comfortable. It's amazing how, after a major purchase, a TIB will sit back and relax, rather than working twice as hard.

Small Items Can Add Up
The large purchases aren't the only symptoms of overspending. Country clubs, association dues, magazine subscriptions, books, clothes, and entertainment can all add up to an overhead cost that may not be affordable. Remember, you're in business for the long haul. It will take hard work and a conservative standard of living to sustain long term success.

Don't Let Your Overhead Price You Above the Competition
Expanding with vehicles, equipment, or unnecessary overhead can force you into a price increase to offset the added expense. This increase could temporarily price you above the competition. A TIB who is conservative with his standard of living and his management of the business will have an immediate advantage over a TIB with a large overhead. His cost of doing business is less and, therefore, he can bid the job lower. His material and labor are about the same, but he can make the same profit and still do the job for less, simply because he has controlled his expenses. A new truck, an office, a secretary, etc., may all be needed and affordable. However, they are often obtained before the workload and operating capital have been built to a level that can support those expenses. Overspending is a real problem. Take it very seriously and keep it under control in your business.

VEHICLES

A pickup truck is usually all that you will require to carry tools and materials to the job site. In most cases, it will not be necessary for you to own a large truck. Learn to utilize the suppliers' services. Most of the time they will deliver materials for a minimal charge. You may see other tradesmen buying big trucks or trailers to carry materials to their jobs, but just because they are overspending doesn't mean you have to.

It's not wise to go out and buy a new truck if you don't already own one. Instead, get a clean, used pickup. Don't believe the hype about new trucks being cheaper in the long run. In most cases the used truck will cost much less even with regular maintenance expenses.

A common excuse for buying a new truck is the desire to impress clients and to look professional. You don't have to drive a new vehicle to impress a client. He probably won't even notice it. Just keep it clean and running. A dirty vehicle is a real turnoff, but usually a new truck won't make much of an impression. The client will be sold on your personality and workmanship, not what you drive. If he makes an issue about your driving a used vehicle, you probably should question his character.

In your business, you must first crawl, then walk, then run. If you overspend on vehicles, it can turn out to be a major loss of needed working capital.

EQUIPMENT

The proper tools and equipment are needed in any trade, and professional grade equipment is a good idea, but a lesser grade and more economical purchase of tools might be wise from time to time. If you feel a purchase is necessary, look carefully and seriously at used equipment. Almost any large machinery can be purchased used at a fraction of what it would cost new. There will always be someone, somewhere, wanting to sell quality tools and equipment at a good price. Watch the classified ads and ask suppliers and other TIBs if they know anyone liquidating their assets to raise cash. Occasionally pawnshops have good deals on used tools.

The decision to purchase more sophisticated and expensive tools and equipment is made only after you have determined that the tools you now own cannot accomplish a professional job in the time required. If what you have can do the job, stay with what you have. Unfortunately, many equipment purchases are a result of being fascinated by and admiring a new toy, or getting a feeling of excitement from the mere ownership of a new tool. Watch out for this tendency and keep it under control. Most TIBs have holes burning in their pockets when they enter a tool supply store.

Rental Versus Buying

On larger equipment, rental will probably be more economical for years. The reason TIBs choose to rent rather than purchase equipment is that it is cost effective for them at that particular moment. The key is to evaluate whether renting is more to your benefit than buying. Here are four criteria that determine when renting will be cost effective:

1. When the piece of equipment will not be used often enough to justify the initial expense and maintenance and storage costs.
2. When scheduling of different jobs requires the same function to be done at the same time, and you don't have enough equipment to do two or more jobs.
3. When out-of-town work is being performed and the cost of transporting equipment is too high.
4. Sometimes there are tax advantages to having a standard rental expense rather than a capital investment in an equipment purchase. Consult your accountant for more information.

Leasing as an Alternative

If you are in need of equipment but can't afford the initial expense, you might consider leasing. In recent years the leasing industry has grown and just about anything can be leased —trucks, automobiles, computers, tools, and equipment of all kinds. In fact, the leasing business has become very competitive and the price and terms of the lease can be negotiated. However, there are both advantages and disadvantages to leasing.

Advantages of Leasing

1. Leasing companies tend to be more flexible in their financing than banks or other lending institutions. Terms and price can usually be negotiated.
2. In some cases tax deductions for the lease cost are more beneficial than depreciation on purchased equipment.
3. Leasing generally requires little down payment leaving the cash reserves available for operating expenses.

4. Leasing companies usually carry warranties on their equipment and will fix or replace it with little cost to the TIB. This will vary from company to company.

5. Short term leases will allow you to upgrade equipment so that it doesn't become obsolete, such as computer equipment. In most cases you can trade in the leased equipment for the newest model available at the end of the lease period.

6. Some long term leases on major equipment are "lease to own" agreements with a percentage of the lease amount applying to the purchase price of the equipment.

Disadvantages of Leasing

1. Leasing equipment usually costs more than purchasing. Even with a bank note a TIB has the opportunity to eventually pay it off. With leasing there will always be the monthly payment.

2. Leasing is a long term commitment. Rental equipment may cost more in the long term, but it can be returned at any time, ending the commitment.

3. If for some reason the monthly cost of the leased equipment is no longer affordable, the TIB may find it difficult to get out of the lease agreement. Selling can be difficult because of the high payoff that comes with leased equipment. In fact, the leasing company usually requires total payment of the lease agreement even if the equipment is returned. Many lease agreements are not transferable, allowing someone else to assume the payments. It can be tough getting out of a lease agreement prior to the end of the lease period.

4. There is often no equity in a lease agreement. Once the lease period is over, the TIB has no ownership of the equipment, and no equity built up. There is nothing to show for your investment.

Purchase, rent, or lease; new or used, look at the options very carefully and make sure your decision is made with good financial judgment based on the actual needs of the business and responsible spending.

WAREHOUSES

Having your own warehouse where materials and equipment can be stored may seem the ideal set up, but for a small business the warehouse can create an unnecessary burden. There are several seemingly valid reasons for having a warehouse. For instance, if you buy in quantity, the material price structure is lower...it provides a place to organize tools and equipment...a place to store vehicles...a place for the employees to meet and get organized...and the list goes on. The use of a warehouse may seem like a good idea, but there are definitely some drawbacks.

Use Your Jobs and Suppliers to Store Materials

First, the pricing structure may be lower if you buy materials in quantity, but the expense of the warehouse, the cash outflow needed to buy in quantity and labor to move the materials two to four times will offset any savings. The best way is to learn to use the supplier's services efficiently by letting them be your warehouse. Have accounts at several suppliers around your area so there will always be one available near the job.

Also, the warehouse will provide a place to collect stacks and piles of leftover materials that may never be used, taking up valuable space. The best policy is to carry leftover materials to the next job. Use every available piece as soon as possible. When it becomes damaged, get rid of it. The labor required to move the same materials time after time will offset the savings of storing those materials.

Use Storage Space Already Available

For materials that do require storage, learn to use space already available. Your garage can be the perfect place, if you do not alarm the neighbors. They will certainly complain that you are in violation of zoning laws if you cause them any inconvenience. Avoid having large delivery trucks come to your house, and don't do a lot of banging and sawing. Be courteous and considerate, and the neighbors will probably work with you.

The warehouse also provides a place to kill time for

both you and your employees, and that's the last thing either of you need. Sure, you need to get organized, but the warehouse can create lots of dead time if not properly supervised.

If you must have some form of warehouse space, few options will prove more suitable than the future goal of a country home with room for a workshop/warehouse. Lower taxes along with being close to the workplace can make your tasks a little more enjoyable. Many days will begin and end at the shop. Being the supply runner for the business can be a time saver. Don't make this a first step. It may take several years before this goal can be realized.

OFFICE ORGANIZATION

Rarely does a TIB in any trade need a full-time office outside the home, and time spent in the office at home must be held to a minimum during the day. A tradesman turned businessman will have to work with tools, clean up, teach skills, estimate, collect money, find new business, and supervise the work. Feet propped on the desk or going to eat long lunches at home and playing with your spouse and kids doesn't get the job done and doesn't make money. Make your vehicle a mobile office. Operate from a briefcase, always carrying the daily planner, calculator, bid forms, workbooks, and any other regularly needed items. The sit-down work can be done in the evenings. The TIB should have an area in the home designated as the business office.

It is common among tradesmen starting out in business to have little or no concept of a normal office setup. Typically, not many tradesmen have had the opportunity to work in or around an office environment.

Don't let it bother you if this has been your experience. The ability to organize and run an office is a learned skill. Over a period of time, you will learn how to operate efficiently in the office.

One of the problems facing the TIB is where to set up the home office. It's not so bad if you have an extra bedroom in your home that can be converted to an office. If space and privacy are limited, use a closet with a large shelf at desktop level. It won't take up much space if you

spend the time to organize the paperwork. A place for everything, and everything in its place is a must. Time spent on paperwork can be cut in half and will be much more enjoyable if everything is properly filed away. Reduce the time you must spend in the office by keeping most of the paperwork current during the day while you are waiting for appointments or when you have a few moments to spare.

The first step in properly organizing an office is to set up a file for every company or individual that supplies you with either services or supplies. This includes material suppliers, subcontractors, utility companies, professional services, etc. Each file should include copies of paid invoices, credit applications, and all correspondence. Use a filing cabinet or desk file drawer to keep these file folders in alphabetical order.

Another set of files for every contract or job should be kept separately. This should include both jobs that you have had in the past and jobs that you are currently working on. All important information about each job must be kept in this file including contracts, proposals, permits, specifications, details, letters, worksheets, notes, etc. Every time material is ordered, a letter is written, or contracts submitted a copy of that transaction must be immediately filed in the proper job file. This procedure will take a little extra discipline, but it will be worth it in the long run. The one time that you decide not to make a copy of a particular document and place it in the contract file is the one time that you will really need it!

Keep It Simple
No system, no matter how well organized, should be so complicated or time consuming that it takes all your time or is confusing. You are the person who the filing system must accommodate, so make it work for you. Simplicity is better 100% of the time. Find out how to record accurately all transactions and agreements without creating a mountain of paperwork. Keep it simple, simple, simple.

Follow Through with the Details
Hopefully, as a tradesman, you don't take shortcuts in the work process. You wouldn't leave a hole in the wall or forget to install window trim. On the job,

details are crucial. In the office, the same attention to details must be paid. The only way to be sure that a detail is remembered is to write it down.

An excellent way to keep up with the details is to carry a daily "diary" in which you record all calls and conversations that occur during the course of a day. Refer to the time management section of this book for information on daily diaries and planning workbooks. With this you can easily document items that might need reviewing later without having to rely on your memory for details that were discussed. In your daily diary record such things as the date each job phase was completed, daily weather conditions, verbal instructions or agreements with the client, and verbal price quotes from suppliers. The ability to look back and recall these vital transactions is immeasurably valuable. This diary can also be reviewed at the end of each day while in your office, to evaluate and take action on necessary items.

Other office-related details such as filing shipping tickets, sending out letters on time, and paying invoices should be completed as quickly as possible. Many don't realize it, but these details are just as important as the work performed on the job. If they are not completed in a timely manner, the mountain of paperwork can quickly become overwhelming and perhaps costly.

The Need for Privacy

An office should be a private place in your home — a place where the family noises can't interfere with your work. Keep all unrelated items away from the work area to avoid unnecessary distractions. For some TIBs phone calls at night will be a regular occurrence, so make sure a phone is near the desk.

It is always a good habit to answer your telephone in the work area, so that you can be near any information pertinent to the conversation and away from family disturbances. You want to convey to the client a feeling of professionalism, and a crying baby or television noises in the background could make him think the reverse.

If your personal telephone doubles for the business phone, it is advantageous to answer the phone by pleasantly saying your name: "Hello, this is John Smith."

This sounds more professional than just saying, "Hello."

It is also important to keep the business separate from the personal for your family's sake. When space is a problem, a complete separation of business and private life may be an impossibility. Unfortunately, this never gives the TIB a sense and quitting and going home from work. After a while this can have a disturbing effect on the family.

What Office Supplies Are Needed

In order for an office to function properly it must be stocked with a few essential supplies. Here is a list of supplies commonly used in an office, all of which can be purchased at any office supply store. This list is not all inclusive, but it's a start:

stapler and staples
staple puller
paper clips
pencils
pencil sharpener
ink pens
file folders (legal and/or letter size)
folder labels (with adhesive back)
rubber bands
correction fluid for typewriters
calculator with tape
postage stamps
scratch pads (for notes and sketches)
eraser
stationery
plain white envelopes (business size)
architectural scale (for reading plans)

A typewriter is a useful tool if you can afford it. Don't buy one of the fancy, expensive machines. Look for ads on a good used machine. The startup costs for office supplies (not including the calculator or typewriter) will run anywhere from $50.00 to $100.00, depending on the quantity you buy.

Send Informal Messages Quickly and Easily

Another aid is the three-part snap-out memo for information correspondence. This type of form is found at the office supply store and is used to make informal information transfers. In other words, when you want to give someone information in writing but don't want to take the time to write a formal letter, this is a very acceptable procedure.

You send two copies to the person to whom the memo is directed and keep one for your records. The person to whom you are writing can return one of the copies with his response. You can use this as a substitute for dictating messages to a secretary, and they can be sent to subcontractors, suppliers, job foremen, etc.

This and other procedures will help speed up the transfer of information from the TIB to other individuals. Basically, business management is a constant transfer of information from one source to another. The efficiency and accuracy of these information transfers is part of the definition of a professional TIB. An office with an organized filing system will promote the efficiency of the business and will eventually make the TIB's job much easier.

SECRETARIES

How easy it would be if you didn't have to do paperwork. Unfortunately, paperwork is a very real and necessary part of running a business.

Hopefully, your spouse is able to help with letters, typing, and record keeping. The tradesman can do the work so much more easily and efficiently when someone else can keep the records in order and the bills paid on time. However, if your spouse is employed full time, leave it that way. You can't afford to get independent too quickly. The object is to grow slowly without a great deal of risk. The record keeping and secretarial work will just have to be done in the evenings.

For those who absolutely must have help with certain secretarial duties, there are many secretarial and bookkeeping services available. Use the yellow pages and talk with several agencies until you find one that suits you and meets your needs. Don't talk big with them. Admit you are small time and in need of only limited services. Be sure that you are comfortable

with the agency you choose. Remember, if the paperwork and record keeping can somehow be handled without hiring one of these agencies, don't spend the money, but do what works for you.

In some cases, a TIB will have a friend or acquaintance with secretarial skills who is looking for extra work outside of their regular daytime job. A couple of hours once or twice a week can be helpful in completing the paperwork and much more economical than hiring a secretarial service.

ANSWERING MACHINES

An effective and economical way to stay in contact with clients is the use of an answering machine. They can be purchased at most discount department stores. Get one that can play back your messages when calling from another phone. This way, you can check the messages at least twice a day from almost anywhere.

Some people are reluctant to leave messages on a machine, and certain clients may have to be contacted regularly in order to keep them happy. However, the average person will leave messages on your machine if he knows that you will call him back immediately. Answering the phone with a live voice is not as important as returning calls within a short period of time. Return every call, no matter who it is and even if you know what the call is about. Give the impression to both your clients and associates that every call is handled immediately. This will go a long way in customer relations.

Don't let a client ever suspect that you are leaving the answering machine on while you are at home in order to screen the calls. Once you're caught screening calls with the answering machine, its usefulness diminishes. The small business is totally service oriented, and you must use the answering machine as professionally as the other tools of your trade. Make yourself available night and day. Offer a personalized service that the large business can't (or won't) provide.

PAGERS

Pagers are a must after completing even a handful of

jobs. Accept the annoyance of these little rascals. The TIB must be available. Promptness in response to the client is of the utmost importance. Pagers can either make a client mad or they can cause him to be impressed with your operation, depending on how promptly you respond to his calls.

There is no need to put one on every employee, just the man in charge, and don't give out the number to everyone you know. This is mainly a tool for the clients. A pager can save you money in labor, gas, and materials when used properly. For example, if your foreman is working on a job thirty miles away and a call comes in to do a repair in his area, another trip might be saved by using his pager.

There are many companies providing a variety of pager services. However, there are basically four types of pagers:

1. Voice Pager — After the tone the caller will have a few seconds to leave a voice message which the TIB hears through the pager. The advantage of a voice pager is that it gives the TIB messages without his having to find a phone and call in for the message. The disadvantage is that if the TIB is around noisy equipment or machinery, the message cannot be heard.
2. Digital Display Pager — Callers must use a push-button phone to leave messages on the digital pager. After the tone the caller will punch in his phone number or whatever number he wants the TIB to call. The advantage to the digital pager is that it will hold three or more numbers indefinitely. This way, if there are loud noises or if the TIB is unable to write down the number immediately, he won't lose the information. The disadvantage to the digital pager is that the TIB has to make a phone call to find out the message. The only information that can be transmitted is a series of numbers.
3. Tone Pager — The tone pager doesn't allow the caller to leave a message or phone number. When the TIB's pager number is dialed it sets off a tone on the pager. With this type of pager the TIB must let only his office or main supervisor

have the number. Otherwise he would never know who called or why. The tone pager just tells the TIB to call his office for a message.
4. Alphanumeric Display Pager — This is similar to the digital display pager except that it allows the callers to send a series of numbers and letters to the TIB. This way the TIB can receive a written message. However, the caller must have special equipment to send the message. This type of system is usually handled by the TIB's office only. The callers leave messages at the office who then send the message to the TIB.

Lease or Purchase

Pagers can be either leased or purchased. However, they are similar to your personal home phone in that if you buy the pager you will still have a monthly charge for use of air time. Prices will vary from area to area and from company to company. If you want to purchase a pager the cost for a voice pager will range from $200.00 to $300.00 and approximately $100.00 to $150.00 for the digital pager. After the purchase, there will be a monthly air time charge of $15.00 to $30.00 per month. This is a set fee and does not vary depending upon the number of calls. Like phones, if the pagers break and need repair, the TIB will be responsible for replacement or repair cost.

Leased pagers are serviced by the rental company, and the TIB is rarely charged for repairs on a damaged pager. Leased pagers cost approximately $25.00 to $45.00 per month depending on the type of pager. This includes air time rental. The TIB will sign a lease agreement for use of the equipment, but it is usually a non-binding agreement which can be terminated at any time by the return of the equipment.

Answering Services

There are also answering services available that will provide the TIB with a pager. As they answer the incoming calls they will page the TIB and give him the messages. This may be an advantage for the TIB who doesn't have someone in his office answering calls. This service usually runs between $50.00 and $75.00 per month. The advantage of using this service versus pagers only is that the client is able to give a live voice the message without worrying about staying near the phone until the TIB calls back. These answering services as well as pager companies are listed in the yellow pages.

MOBILE TELEPHONES

With today's technological advances the mobile telephone is becoming much more affordable to the general public. However, most TIBs will not have the need for a mobile phone in the early years of their business. It is something to consider much later as the business grows, as the need presents itself, and as it can be easily afforded.

The early models of mobile phones were quite expensive. They had poor reception and there were a limited number of channels available. It was sometimes very difficult to make or receive calls. But with the development of cellular phones those problems have become a thing of the past, and the cost of the equipment has dropped considerably.

The equipment can be either leased or purchased. If you purchase the phone, the prices will vary drastically depending on the type of equipment and phone company used. The prices will range from $500.00 to $2,500.00 which will include the installation in a vehicle. Phones are like cars; there are Chevrolets and there are Cadillacs, each with a variety of features such as built-in phone directory and automatic redial. Buying a phone should be done as carefully as buying any other piece of equipment. In addition to the purchase, there is a monthly fee for air time. This charge will vary depending on the amount of air time used. The phone service will document every second the phone is being used on a call and charge accordingly. This includes both incoming and outgoing calls. Some TIBs don't realized the amount of monthly expense that can result from owning a mobile phone, particularly when the phone is new. Especially if the number is given out to several people, he may receive numerous calls from other individuals, all of which he has to pay for. It's incredibly easy to charge $100.00 to $300.00 per month on mobile phone calls.

Leasing mobile phones is another option with the monthly rental fee ranging from $20.00 to $50.00. There is usually a substantial deposit required when signing an agreement. The air time fee will be billed in addition to this rental fee and will be structured as explained above.

Many time mobile phones are purchased before they are needed or can be afforded. The mobile phone is not just something to add to your prestige, but a tool that makes business function efficiently. If the cost is greater than the need, avoid the purchase even if you can afford it.

COMPUTERS

High Technology...sounds great, doesn't it? The computer age is upon us. It has been said many times that a business without a computer cannot compete in today's market. This is simply not true. Yes, a computer is a helpful tool, but a TIB running a small business can function very well without the use of a computer.

Record-keeping procedures, bids and estimates, letter writing and cost analysis can all be done by hand, efficiently and effectively, in a small business setting. In fact, secretarial and accounting services may be able to produce computer-generated work more economically than if you purchased your own unit.

Along with the computer age comes the high tech salesman. He can make that computer look like the answer to all your current and future needs. He will have you believing that you can't go a day longer without owning their moneymaking/saving machine. Unfortunately, many salesmen don't care whether you can afford it or if you need it. Buyer beware. Yes, many companies are operating very successfully with this equipment. It's a great tool...but is it right for you and your business? Ask this question before purchasing a computer: "Can I operate professionally and efficiently with less expensive equipment or can I operate without it all together?" If the answer is yes, don't buy it. It's that simple.

Now that we've established priorities, let's discuss

the use of the computer in a TIB's business. As a TIB's business grows, the need for a computer may arise. It's very important, however, to know two things before buying the computer. "What are the specific needs of my business?" and "What software and hardware are available to meet those needs?" Too many TIBs rush out to the computer store and buy whatever looks good. There are hundreds of options, all of which tend to frustrate the unskilled buyer. The more you learn about what is available, the more confused you become about what to buy.

First, let's explain the difference between hardware and software. Hardware is basically just what it sounds like. The computer consists of four basic hardware parts: monitor (looks like a TV screen), disk drive (the main workings of the computer brain), key pad (like typewriter keys), and the printer. When choosing hardware there are several things to consider, such as memory capacity of the computer and the printing quality of the printer. These are not easy decisions for the untrained buyer.

Software on the other hand is what makes all of the hardware work. Software is the computer programs that make the computer carry out certain functions. In a TIB's business the software is the most critical element. Therefore, it is best to choose the software package before buying the computer, because not all software works on all computer hardware. The reason this is so important is because of the accounting requirements of a construction-related business. Construction is very different than most other businesses and requires very specific software to handle the job of cost reporting. However, there are a variety of software manufacturers catering to the construction industry. Refer to the Resource Materials section for a list of computer reference materials.

With a computer and the right software, a TIB can write letters, organize data, keep books/accounting, bid jobs, and write proposals. The use of the computer can be a great advantage. But remember this. Don't buy anything before you know how to use it. It is a useless piece of junk unless you can make it work for you. The TIB needs a tool, not an expensive desktop ornament. Computers can be terrific machines in today's business world, or they can be the biggest problem you ever had, so be informed before making a decision.

3. TECHNICAL SKILLS

QUALITY WORKMANSHIP

The premier sales tool of a TIB is the quality of work. Quality isn't difficult to obtain from time to time. Maintaining a consistent level of quality from job to job is what is tough. It requires daily on-the-job supervision by the TIB with his tools on, training and setting an example. Sometimes the cloud of paperwork and a heavy workload can cause you to lose that eye for detail. Be your own company's worst critic. Misleading yourself by pretending not to see flaws or mistakes will be costly and damaging to your reputation, your self confidence, and your future.

Quality, in the eyes of the client, will be reflected in areas other than workmanship. Promptness, following instructions, and fulfilling every obligation are equally important. The way in which your proposals, billing, and change orders are handled will convey to the client your ability to operate in a professional manner. Remember, an eye for detail should be applied to every aspect of your business from appearance to actions. It is important to give each client every assurance that the job is under control and the quality of workmanship will be consistent, with or without his presence on the job.

The reality of your commitment to customer satisfaction may have its greatest challenge when called upon to do warranty work. It is important to stand behind every job. For instance, suppose you have received final payment on a job completed over six months ago, and late one rainy afternoon the client calls complaining of a serious leak. Do you wait until tomorrow when the weather is clear or when you have more time, or do you send someone over right away?

Showing concern promptly will eliminate most irritations, and it's cheaper than lawsuits and lengthy phone conversations. There is no sales tool greater than taking care of warranty work quickly and efficiently!

Tradesmen who do poor work are a dime a dozen. Be a TIB who sets an example. Choose a level of excellence in your work. It takes more time and effort to be the best, and you should strive to establish your reputation not only as a superior craftsman, but also as a qualified businessman. When established, your business will go the distance, while others that seemed to have a measure of success will eventually fall out of the race.

BLUEPRINT READING

It will be difficult either to bid accurately or to complete a job without a good working knowledge and ability to read blueprints. Hopefully, during your training as a tradesman, plan reading was a regular part of your responsibilities. However, you may be able to get by with just a limited knowledge of architectural plans. Spend as much time as necessary with a set of plans so you understand the details relating to your bid.

If you have not had the opportunity to work with a variety of plans, there are trade schools and community colleges that provide night classes in reading blueprints. These courses are an excellent way to learn this skill. The more you know about every section of the plans (architectural, structural, mechanical, plumbing, electrical, etc.), the greater the chance of eliminating costly errors. Refer to the Resource

Materials section for a list of materials that teach how to read blueprints.

Architects and engineers will show a detail of the building in the strangest places. Sometimes it almost seems like a game to see if you can find the hidden details. And don't assume that the plans are without error. Mistakes can be found time and time again. That's why it is important to research carefully every detail of every section on every page. From time to time you will find a major mistake or discrepancy and won't be able to resolve the problem yourself. At those times it may be necessary to make a phone call to either the architect or the engineer for clarification.

Architect versus Engineer

First, let's explain the difference between architects and engineers. When a client needs a set of design drawings for a new project, he will usually make the initial contact with an architect. The architect, under the guidance of the client, will be responsible for designing the layout of the building. He will determine how it will look, how large, what materials to use, which colors, the room finishes, etc. On small buildings, particularly residential work, the architect will also design the electrical, plumbing, and HVAC (heating, ventilating, and air conditioning). This usually amounts to little more than determining the location of fixtures, based on local codes and client needs.

If the building is relatively large, the architect or owner will hire engineers to help design the structural, mechanical, and electrical portions of the building. The engineer's job is to design a structure that will hold up and make operational this beautiful monument that the architect has designed. If the structural soundness of a building is in question, the first step is to check out the engineer's design. The engineer will be an important figure in the career of a TIB.

Handling Problems and Changes in Plans

When finding problems on the plans, as previously stated, the ability to understand how the plans are organized will help you determine who to call to get the right answers. If the problem is related to room fin-

ishes, floor plans, doors, windows, etc., the architect should be contacted. Many sets of plans are broken down into sections that describe the type of details on each drawing. For instance, drawings A1, A2, A3 are Architectural relating to floor plans, building elevators, room finishes, etc. The architect will be directly responsible for these details. Drawings S1, S2, S3 are Structural relating to foundation details, structural details, framing, connections, etc., which are the responsibility of the structural engineer. Drawings M1, M2, M3 are Mechanical relating to HVAC, venting systems, exhaust systems, etc. These drawings sometimes include plumbing details. However, plumbing drawings are many times shown as P1, P2, P3. The civil engineer will be responsible for these details. Drawings E1, E2, E3 are Electrical relating to electrical details, which are the responsibility of the electrical engineer.

In addition, on some larger jobs, a set of specifications will accompany the plans. These specs will be divided into several sections and will identify products, materials, and methods of application. In most cases these specs are standard from job to job. Just find the section that deals with your trade and gather the necessary information.

With this information you will be able to determine how to approach and handle a design problem. There is, however, an unwritten code of ethics that requires any subcontractor to channel all inquiries through the general contractor until approval has been given to have direct contact with the architect and/or the engineer.

Architects can be very sensitive about their work, and your approach should be made cautiously. Typically, architects don't like to be proven wrong. If you get on their bad side during the initial contact, it will be down hill from there.

Engineers, on the other hand, generally have a greater ability to relate to the technical problems of the tradesman. They may, however, relate in the same fashion as the architect when it comes to their design. This is caused by the tradesman's inability to understand the delicate balance of technical information that is taken into consideration during the design process, and an engineer's inability to see the practical application of his design and the very real problems in that application that only a tradesman can understand.

Their protectiveness of the design is somewhat understandable as they must put their professional engineer's seal on each drawing, giving them a measure of responsibility for the soundness of the design. In any case, your attitude will weigh as much or more than your knowledge of construction when relating to either architect or engineer.

Document Plan Changes

Now that you've determined whom to talk with, called about the problem, and clarified the situation, follow up with a letter. This will provide you with a record of the call and the solution to the problem. At the very least, document the call in your diary. The client may issue an addendum or change order to the contract, but in many cases your notes will be the only record you have of the change. There will be countless calls, changes, and clarifications every day, and it is impossible for anyone to keep up with all of it efficiently without writing the information down in an organized manner.

Hiring the Engineer

It may be necessary on occasion for a TIB to hire the services of an engineer for a particular job. Let's say you have been given responsibility to help design and build a certain project. You may need to place a foundation on questionable soil conditions that must withstand the movement stress caused by bad soil conditions. A structural engineer will be needed to assist in the design of the foundation. Or perhaps in the process of remodeling a factory you need to add heavy equipment to an already overloaded electrical service. An electrical engineer can check out the design to establish what needs to be constructed.

After receiving the design from the engineer, whether you agree with it or not, it is your responsibility to build it exactly as it is designed. Varying from the design will put you at risk of being responsible for design integrity.

Taking unnecessary risks can create long term disasters.

Very seldom will a TIB need an engineer because most projects are pre-engineered, and the TIB's bid is based on this predetermined design. However, an occasional potential job surfaces that requires additional effort and creativity by incorporating the skill of the TIB with the services of an engineer. Utilizing these engineering services can many times secure a few extra projects.

Be sure always to have the engineer document his design and seal the document or drawing with his engineer's seal. Get an original seal and signature from the engineer and keep it on file. You must have a record of his design clearly showing his legal record of the design. If something goes wrong, it may happen years from the time the project was completed. Careful documentation can keep the TIB out of serious trouble.

When hiring an engineer, there are a few things to remember. Don't assume that all engineering prices or abilities are the same. Some engineering firms will charge more than others, and some are not interested in doing small engineering jobs in a specialized field. Most engineering firms are more proficient in certain areas but not as knowledgeable in others. Before making a decision, call around and ask other TIBs and individuals in your field to recommend a qualified engineer.

KNOWLEDGE OF MATERIALS

If you are an experienced tradesman, you probably have a good working knowledge of the materials of your trade. However, with today's technology, there are new products available every day. Making yourself aware of what's available and how it is used is critical if you want to keep up with the competition.

When bidding a job, the TIB will usually be supplied with a set of architectural plans and specifications. Certain materials, brands, or methods of application may be designated. Read these documents very carefully. Mistakes can become costly when your bid includes a different or less expensive product than that specified in the plans. For example, if you bid No. 3 studs for a building that called for 1500 psi lumber, you will be in serious trouble. Obviously, in this case, a much stronger grade of stud was specified. Suddenly you may be faced with removing and replacing the inferior lumber.

Local and state codes will sometimes dictate types and strengths of materials to use. Detailed research should be done to determine how and what to bid.

If you are doing work in a certain municipality or group of municipalities, become as familiar as possible with the codes in those areas. There will be situations when you won't have time to research the code books before submitting a bid. Becoming well versed in the codes in advance will help in bidding as well as city inspections. Refer to the Building Codes section for details. From time to time the client will look to you for advice on different materials and products. Familiarize yourself as much as possible with all of the products and their applications. The suppliers should have technical information on all available materials, and it will be helpful for you to keep a file on every type of product used in your field. Take advantage of the many widely distributed reference books, brochures, and samples to help keep you up to date.

Sweets Catalogs

There are companies that have made available to TIBs sets of catalogs which are a group of brochures from their product manufacturers. These catalogs can be obtained from lumber companies and related building supply companies. However, the best resource for product catalogs is the Sweets Catalogs published by the McGraw-Hill Informations Systems Company. The Sweets Catalogs have over fifty volumes containing more than 47,000 pages of product data. Other companies have similar catalogs, but Sweets is the best available that specifically caters to the construction industry.

The Sweets Catalogs can be obtained directly from McGraw-Hill, but are not available to everyone. The product manufacturers have made the catalogs available to qualified contractors. The TIB may or may not qualify

depending on his annual workload and type of work. There are different types of catalogs available, but only two relate to the TIB — the "General Building" and "Homebuilding and Remodeling" catalogs. The Sweets may not be easy to get, but it is well worth the trouble.

PURCHASING MATERIALS

Sometimes having purchasing skills is taken for granted. Shrewd purchasing can produce large profits for the TIB. Initially the TIB will order all materials, but as the business grows it may be necessary to assign this task to an employee. You will want someone who has organizational skills and is good with details, has an authoritative personality, but works well with others. Certainly you must have a person who can be trusted completely, not tempted by offers of "kickbacks."

The TIB should have only one person in charge of purchasing materials. Otherwise there will be confusion in ordering, resulting in duplicate orders, overcharges, and lack of continuity. Many suppliers keep an account card on file for each of their customers. This card lists vital statistics on the TIB's business such as whether you use purchase orders and which individuals in your company are allowed to purchase materials. If the suppliers don't protect the TIBs in this manner, anyone, including past employees, could charge materials in your name. The TIB should give the suppliers only the names of persons vital to the TIB's business. Only these individuals are allowed to purchase materials from that supplier.

Inventory

Before purchasing any materials a TIB must evaluate whether he has material in inventory that can be used. In most cases inventories are a money pit that seems to get deeper and deeper. Thousands of dollars can be wasted by storing materials that never get used and are eventually damaged. The way to resolve this problem is by taking inventory of all available material. This can be done by using the Inventory Sheet (Figure 1). If possible, show the current value of the material so that your inventory can be used not only in the purchasing process but also in the accounting functions (determining total asset value).

If the wrong material is ordered or if several pieces are left over after a job is completed, many TIBs just put it in inventory hoping to use it on another job. Most of the time that material will be shoved to the back of the storage room and forgotten or damaged, yet the TIB had to pay good money for it. If the extra material was the mistake of the supplier he will take it back without question, but if it was a mistake in ordering, returning may result in a restocking charge. Even if this is the case, it may be better to pay a restocking charge than to store something that will never be used.

Purchasing Priorities

The TIB or his purchasing agent must take into consideration several details when ordering materials and choosing which supplier to use. The following are factors that must be taken into consideration:

1. Quantity — The TIB must decide how much to order. Large quantities of items sometimes bring a better price, but this is a waste if the material is not used on a timely basis. Double check quantities before ordering because inaccurate quantities are the cause of most purchasing mistakes.

2. Quality — Try to get the best possible goods for the job. This does not mean that you should order the very best, which is generally more expensive, but that you should order the best for your needs. Ordering better materials than are needed adds needless costs, yet ordering inferior quality materials can result in inferior finished products and wind up costing much more to redo what has already been done.

3. Inventory — As discussed, careful control should be kept of your inventories. It costs money to store materials and every effort should be made to use these goods as quickly as possible.

4. Rate of Use — Knowing how often and how fast an item will be used will determine how much to order. If the item will stop production if you run out, it may be necessary to keep extra on hand. Of course, as work increases or decreases, it is sometimes hard to evaluate the rate of use.

5. Storage Space Available — The amount of space available on the job will greatly influence the quantity to buy. If additional storage space has to be rented to keep materials, it is probably unwise to stockpile.

6. Type and Use of Material — On special items that are revised on a regular basis making the older models obsolete, orders must be very conservative. Also, an item that is rarely used or that deteriorates with time must be purchased with great care.

7. Delivery Costs — Transportation costs for several units are considerably less per unit than the shipping cost for only one unit. The larger the load, the less it costs per unit.

8. Shipping Time — If items are ordered and have to be sent from out-of-town sources, the time involved may be a factor. For those items the TIB must make sure the order is placed in advance so as to allow for the shipping time.

9. Price — Obviously price is not the only concern, but it is one of your major concerns. Getting the best price is certainly an important factor, but haggling or dickering over prices is generally not considered good business practice. Normally, the sales people of the supplier don't have the authority to make price concessions, so the best policy is to accept a quotation as being "firm." However, suppliers will lower their prices when bidding on larger jobs knowing they will receive a larger order. The sales rep for a company will have authority to lower the prices when bidding on large orders. Some suppliers, on the other hand, have different price levels for their customers depending on the amount of business they receive and the relationship with the customer. This method of pricing gives incentive to customers to develop long term business relationships, resulting in better price structure. A TIB must insist on being placed at the lowest possible pricing level, and should make sure this is the case before ordering any materials from a supplier.

10. Service — It is important to become familiar with the type of service received from the supplier. Does he deliver his goods at the promised time? Does he stand behind his product and have competent maintenance service? Does he take

care in packing and shipping to avoid loss and breakage? How does he handle damage claims? How does he process returned materials? Some TIBs have found that better service is worth paying for.

Keeping Accurate Purchase Records

In an effort to increase the profits, many TIBs make a job budget which consists of all of the materials and subcontract labor to be purchased on that job. This can be done by using the Purchasing Worksheet (Figure 2). All materials to be purchased can be accumulated and listed on the worksheet. The quantity, description of material, and estimated unit prices can be shown, creating what is known as the job budget. In other words, the job budget gives the purchaser a guide to use in the buying process. The goal for any good purchasing agent is to try to purchase below the estimated price, thus increasing the job profit by lowering the costs. As the materials are purchased, the actual cost is noted and the name of the company identified, so that the TIB has a quick reference as to who is supplying what. As soon as the order is made, the date is placed in the appropriate column indicating that the transaction was made. Later, when trying to remember if the material was ordered, the TIB will have a record of the purchase.

One very good practice is the use of a Purchase Order (Figure 3). Purchase orders are seldom used by many TIBs but are a vital business tool that offers a measure of protection from costly errors. As materials are ordered a copy of the purchase order should be sent to the supplier. The order should list in detail the items being purchased. The purchase order does two very important things. It eliminates mistakes when ordering by phone, as the supplier has a copy of the order. When an order is sent out with the wrong material or quantity error and a purchase order was not used there is no way to determine who was at fault, the TIB or the supplier. It's your word against his. The purchase order ends all arguments.

In addition, the purchase order eliminates overbilling because the TIB has kept a copy of the order in his file. As the invoice is received from the supplier it should be compared with the purchase order. The material description, quantity, and unit price can be easily verified. TIBs are many times robbed blind because they don't check the supplier's billing. They just pay the bills month after month. The supplier may not intentionally put the wrong information on the bill, but mistakes happen, sometimes quite often.

Check the Shipment

A must for all TIBs and their employees is to thoroughly check a shipment of materials when it is received. The best time to check the order is before the driver leaves the job site. If the material is signed for without comparing the shipment with the packing list/delivery ticket, it will be difficult to prove that the material was wrong or that there were some pieces missing. Many TIBs telephone a supplier and have materials delivered to a job site without ever checking the delivery. As the job is being completed and the workers run short of materials, the TIB doesn't know whether the material was shipped wrong, if he estimated incorrectly, or if material was stolen off the job site. It doesn't take that much time to meet the supplier trucks at the job site, and the effort could save the TIB a substantial amount of money.

JOB SITE MANAGEMENT

The manner in which a TIB starts a job, schedules a job, and works with the other contractors on the job creates a lasting impression on the client and other interested observers. The TIB who makes a practice of starting his jobs promptly usually enjoys the fullest cooperation from the client, and also establishes a good job morale which is critical to a successful project. Prompt starting of a job will bring better business relations with the client and a better reputation, as it is usually the trademark of a successful contractor.

One of the first moves the TIB must make in starting a job is to set up the job organization. Every job needs competent supervision, and if the TIB is too busy to do this himself, he should appoint a qualified employee to do it for him. The TIB should be in regular contact with the supervisor making sure that things are running well. Even with a supervisor the TIB must regularly check the jobs himself to make sure that his instructions are being carried out properly and that the quality of workmanship remains at a high level.

Establish a set time to get together with your employees to discuss problems on the job. To solve immediate problems use pagers to create a communication link between the TIB and the supervisor. To eliminate misunderstood communications, write instructions down in a clear and concise manner, and if the job site is noisy take the supervisor or other employees away from the noise when giving them instructions. Using these communication techniques can resolve many of the regular problems that arise.

Proper job site management requires an understanding of the requirements of the many other trades working around you. All too often one TIB doesn't know what the others are doing, and unnecessary interference occurs. Making an effort to get to know the other people on the job site and their plans can save everyone a lot of headaches. These contacts can be helpful at other jobs also. A familiar face can do wonders for productivity.

An unnecessary problem occurs when the workers run out of materials. The TIB should make sure the material deliveries keep pace with the work. He should also make sure the crews are ready to start work each day with the proper materials, tools, supervision, instructions, and number of people. Too many workers on the job can be a waste, and too few can limit productivity. Scheduling requires planning ahead. Waiting until the last minute to schedule the work creates nothing but problems. Plan for each day ahead of time so that everything will be ready when the workers arrive each morning.

Daily Report

Many TIBs are affected by weather delays, client delays, material shortage delays, etc. These delays can cause trouble for the TIB because it prevents him from completing the project on time. The TIB may not have been at fault for the delay, but he may receive all of the blame. To avoid this problem, the TIB should use his time management scheduling book (refer to the Time Management section for details) to record the daily activities. Note such things as weather conditions, work completed, reasons for delays, when materials are received, changes on the

job resulting in a delay, etc. These brief notes in your daily diary will provide a history of the job and can actually be admitted in a court of law as evidence.

Work Schedule

In an effort to better organize and schedule the workers a Work Schedule (Figure 4) should be used. With a work schedule, you will be able to schedule the available workers in advance (making adjustments for delays). Hopefully, this form will help eliminate overbooking of jobs and will help in the future for estimating similar jobs. With organization and planning you will be able to have as many projects going as you feel you can control. With the work schedule a TIB can see at a glance how well a certain job is progressing.

Job site management can take all of the organizational skills a TIB can muster. Sometimes it is much more difficult to keep things straight than it actually is to do the work. However, with hard work the management skills can be developed and mastered.

BUILDING CODES

Most construction work today is governed by a set of rules, standards, and regulations known as building codes. A building code is a legal document used by state and local building inspection departments as a means of controlling and inspecting the construction process. The use of building codes is the government's way of protecting public health and safety. Losing sight of these basic intentions may cause the TIB to view the codes as troublesome and a nuisance. Understanding the importance of the building code and the government's desire to protect the public will help the TIB to work within the system and cause his jobs to run more smoothly.

Understandably, some TIBs will not have much involvement with the local building code because the nature of their work does not warrant strict control. For example, wallpaper hangers, carpet layers, and painters will not be as critically governed as electricians, plumbers, carpenters, and of course, TIBs involved in general construction. In addition, many rural areas are free from government control where the county government has not established a building inspection department.

At first glance, the code book may seem complicated and overwhelming. As a result, some TIBs ignore the code book and rely just on their experience. Obviously if a TIB has been working in the same area for several years he will learn much of the code simply by experience and trial and error. Unfortunately, those errors can be very costly. Therefore, the TIB must accept and use the local building codes. He must determine in his own mind that the applicable codes are for his use and in his best interest.

The Different Types of Building Codes

Sometimes it seems that every city in the United States has its own building code with no two cities alike. Except for a few large metropolitan areas which have written their own codes, most cities have selected, adopted, and revised an established model code. Almost every city has adopted one of only three model codes. Then they revise, change, add to, rework, and reorganize it. This is done over a period of many years as they attempt to address problems in their city. That is why it seems as though every city has a completely different code. The three model codes that the states and city governments use as a base are published by large membership organizations, and each includes a building code, a plumbing code, a mechanical code, a housing code, a fire prevention code, and other related documents.

The three major model codes are the Uniform Code, the National Code, and the Standard Code. Each has a long list of available publications and are as follows:

1. UNIFORM CODES — Published by the International Conference of Building Officials (ICBO), 5360 South Workman Mill Road, Whittier, CA 90601. ICBO publications include:

Uniform Building Code
Uniform Building Code Standards
Uniform Mechanical Code
Uniform Plumbing Code
Uniform Housing Code

Uniform Code for the Abatement of Dangerous Buildings
Uniform Security Code
Uniform Administrative Code
Uniform Sign Code
Uniform Code for Solar Energy Installations
Uniform Code for Building Conservation
Dwelling Construction Under the Uniform Building Code
Uniform Fire Code

2. NATIONAL CODES — Published by the Building Officials and Code Administrators International (BOCA), 4051 West Flossmoor Road, Country Club Hills, IL 60477. BOCA publications include:

National Building Code
National Mechanical Code
National Plumbing Code
National Fire Prevention Code
National Existing Structures Code
National Energy Conservation Code

3. STANDARD CODES (Formerly the Southern Standard Building Code), published by Southern Building Code Congress International, Inc. (SBCCI), 900 Montclair Road, Birmingham, AL 35213-1206. SBCCI publications include:

Standard Building Code
Standard Plumbing Code
Standard Mechanical Code
Standard Gas Code
Standard Fire Prevention Code
Standard Housing Code
Standard Swimming Pool Code
Standard Unsafe Building Abatement Code
Standard Amusement Device Code
Other SBCCI Code-related publications include:
Standards for Installation of Roof Coverings
Standards for Existing High Rise
    Group R — Residential
    Group B — Business Buildings
Standards for Flood Plain Management
Standard for Noise Abatement and Control

One of these three major model codes has been adopted by each state or city government. Interestingly, the government acceptance of the codes has clearly defined boundaries. The influence of each model code organization has divided the country in basically three sections. Each state (shown on Figure 5) has accepted one of these codes as its statewide model code. Each city government has the responsibility of making it work for them by revising and adding to it as necessary.

In addition to the Uniform Codes, National Codes, and Standard Codes there are other codes that have been widely used and accepted by most cities and states. The National Electric Code (see below) in particular is accepted in most areas as the electrical standard. These codes are used in the same way as the code books previously listed. Each addresses specific areas in construction.

1. NATIONAL ELECTRIC CODE — Published by the National Fire Protection Association, Batterymarch Park, Quincy, MA 02269-9990.

2. ONE AND TWO FAMILY DWELLING CODE

Approximate Areas of Model Code Influence

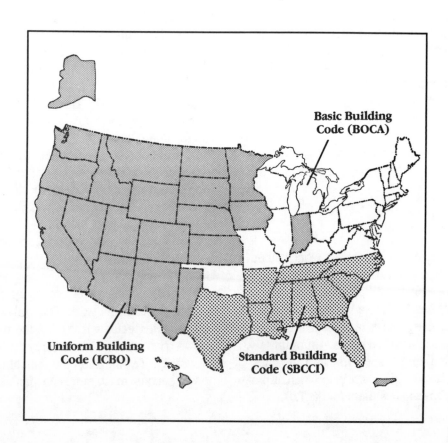

— Published by Council of American Building Officials, 5203 Leesburg Pike, Suite 708, Falls Church, VA 22041.

3. MODEL ENERGY CODE — Published by Council of American Building Officials, 5203 Leesburg Pike, Suite 708, Falls Church, VA 22041.

It's important to remember that while these codes are used as a base on which a city will build its code requirements, it is necessary for each TIB to obtain from that city a copy of the code with all of the revisions and additions. Usually the local bookstore or the city building department has a copy or can help you find a copy. Each of the model codes is updated about every three years and is referred to by the year of revision such as: UMC 85, NBL 87, NEC 84. When purchasing copies of a particular code book, be sure that the year of publication is the same as that required by the city or state.

Most TIBs won't need a full set of all of the available codes. In other words, a plumber might purchase the Uniform Plumbing Code and possible the Uniform Building Code, but he won't have much use for the National Electric Code or Uniform Mechanical Code. Contact the local building inspection department for more information on code requirements.

For more information on available reference materials, refer to the Resource Materials section at the end of this book.

INSPECTORS

Few trades go through their process of work without dealing with either state or local inspectors. Some building inspectors are excellent at their jobs. They have a good knowledge of the applicable building codes, and years of experience have given them a practical and common sense application of those codes. Unfortunately, there are many building inspectors who are inexperienced and lack knowledge. These inspectors can produce major headaches and heartburn for the TIB.

Regardless of their experience or their incompetence, these building inspectors have a great deal of power and authority, and they won't hesitate to use it. Your best defense is to become extremely familiar with the local codes.

From time to time, you will be required to change something that you know is not right according to the codes. Before taking action you must first determine if it is easier and less costly to make the change or to "fight City Hall." In disputing an inspector's ruling, you may find it necessary to go to a higher level in the inspection department to determine the accuracy of the inspector's code interpretation. Be careful, however, not to step on anyone's toes. You don't want an angry inspector making your life miserable on other inspections.

When going into a new area with an unfamiliar inspector, it's a good idea to give him a call prior to completing the work. Ask if there are any requirements to which you should pay special attention. It's amazing how most inspectors have "pet items" about which they are concerned. Finding out these details in advance can save a lot of time down the road.

Make Them Feel Special

When an inspector comes on the job, stop whatever you are doing and make him the center of your attention. Be courteous, respectful, thoughtful, and friendly. His job creates a lot of headaches for a lot of people, and the TIB at the inspector's previous stop may have put him in a bad mood. Be prepared, in case he wants to take it out on you. A kind word from you can make his day (and yours) much more enjoyable.

Good working relationships can be developed with certain inspectors which will eventually give them a level of confidence in your work. This will help you immeasurably. An inspector will be much more strict with an unknown contractor, mainly to show the TIB who is boss. Eventually, if you exhibit the right ability and attitude, he will become much more cooperative.

SECTION II
DOLLARS AND SENSE

4. ESTIMATING AND BIDDING

One of the TIBs greatest concerns is in the area of correctly estimating and pricing a job. Unfortunately, many jobs are not bid properly or are not submitted on time. By following a few simple procedures you will be able to estimate like a professional. Here is a list of elements that make up a good estimate:

1. Careful study of the plans for exact quantities of materials.
2. Accurate pricing of materials.
3. Determining projected labor hours.
4. Establishing accurate hourly labor costs.
5. Obtaining dependable and complete subcontractor bids (if required).
6. Allowing for special considerations, such as site conditions and seasonal conditions.
7. Allowances to cover taxes and insurance.
8. Provisions for permitting and license costs.
9. Including adequate overhead expenses.
10. Adding a reasonable profit.

An accurate estimate doesn't automatically guarantee the TIB will get the job because other TIBs may submit lower bids feeling they can do the work at a lower cost. However, a TIB knows where he stands with a well-prepared estimate. He can negotiate with confidence when he has accurately and thoroughly identified the costs.

There are three basic types of bidding: FIXED PRICE, COST PLUS, and UNIT PRICE. Each type is easy to understand if you take it one step at a time. Use the method which best adapts to your trade and is most comfortable for you.

FIXED PRICE BIDDING
Fixed price bidding, simply put, is determining what materials and labor are required to complete a particular job. After adding the necessary overhead and profit to the material, equipment, and labor costs, a fixed price bid is then submitted to the client. Regardless of what it may eventually cost the TIB to complete the agreed upon scope of work, the client will pay only the fixed contract price. This form of bidding requires much more preparation and research, but it is the most common form of bidding a job.

The key to successful fixed price bidding is detailed organization. Using the Bid Worksheet (Figure 6) you will be able to organize and consolidate the bid information.

DESCRIPTION: This column should list all the various categories that are commonly used on a job in your trade. It's easy to overlook an item or two, so to avoid missing items you might want to standardize the bid worksheet. Make your own personalized form by typing out a sample list of items on the bid worksheet form. It might help if you go to some of the job sites as you make a list of these items. There may be too many to list for some TIBs, but at least list the items that regularly occur. If more room is needed, use two or three separate copies of the bid worksheet showing the recap information only on the last sheet.

This personalized list will eliminate your forgetting to include a certain item in the bid because the pre-printed form will list all of the commonly used categories. By using the Bid Book you will always have access to all other prices and bids at your fingertips while bidding other jobs.

Be sure to leave blank spaces at the bottom of the

form for special additions to the list; items that are unique to a particular job as noted in the plans and specifications. Take plenty of time to study the plans to determine the additional categories needed.

QUANTITY: This column will help you identify how many pieces of each item will be needed. Not every category in the description column will require a number in the quantity column, such as a fixed bid from a subcontractor. However, be as specific and detailed as possible so that mistakes from wrong quantity entries are avoided. Carefully examine the plans and make a very thorough takeoff.

UNIT: This column identifies the type of unit that is being counted, such as square feet (SF), cubic yards (CY), linear feet (LF), board feet (BF), pieces (PCS), etc.

UNIT PRICE: This column lists the current price for each item in the description column. It is important to regularly check and verify (at least once a month) the prices per unit for all commonly purchased materials. On larger jobs you can ask the supplier to quote a price for specific materials for those jobs. He will lock in his prices for a agreed period of time thereby protecting you from any price increases from the time you bid the job to the time you buy the materials.

The suppliers will price different types of products in different ways. For instance, lumber is in board feet typically (but watch for bids in linear feet); drywall is in square feet or per certain size sheet; bricks are priced per 1,000; sand and concrete are priced per cubic yard; and doors and windows are priced per each unit used. Become familiar with your materials and how they are priced.

MATERIAL COST: This column is the total material unit price for each item in the description column. It is obtained by multiplying the number of units by the individual unit price. If sales tax is applicable to your business purchases, be sure to add it to the total material price for each item. Freight may be applicable and should also be added into the total price for each item. By not adding these extra costs to the totals, you may leave out hundreds, if not thousands, of dollars of

costs in the bid price. In some cases, such as subcontractor bids, you will not use the quantity or unit price column. In this case, place the total price for that item in the Material Cost column.

LABOR COSTS: There are two ways to estimate labor costs. One is to determine the cost of labor for each separate major category on the Bid Worksheet, such as framing, drywall, painting, cleanup, etc. Just take your total man hours and multiply them by the average gross cost per hour for each laborer (salary, plus taxes and insurance, plus fringe benefits). Be sure to include all of the hidden costs when establishing the cost per hour, not just the amount on the paycheck.

The second way to estimate labor costs is to estimate what the total hours will be for the total job without separating them out for each category. Place that total in the space provided on the Bid Worksheet as being the total labor cost. Do what works best for you.

EQUIPMENT: This column should include owned and rental equipment used to complete the job. Owned equipment would be bid at a figure covering the cost of ownership (purchase price prorated over the life of the equipment, taxes, insurance, fuel, maintenance, etc.). For instance, you might include in the bid a charge of $25.00 a day for a pickup truck and hand tools.

Rental equipment would be charged to the job at actual rental cost plus cost for delivery and return. Rental equipment may be required as a one-time rental for that job only; but owned equipment is also a very real cost and must be charged to the job, even if there won't be a visible expense during the course of the job.

Recap the Bid Worksheet
MISCELLANEOUS, OVERRUNS: There will always be miscellaneous items and material waste that may never be identified and counted except by a method such as adding a miscellaneous percentage to the total material and labor costs. This figure will vary depending on the geographical area, economic conditions, and size of job. However, it will usually range between 2% and 8%.

On large jobs or on non-waste items, this figure will be lower than on smaller, more complicated work. However, don't let your client know about this method of calculating the bid. He will interpret this figure as profit and may try to get you to take it out. This is a direct cost item and should be treated as such.

OVERHEAD: This item includes costs not directly affected by the jobs, such as supervision salary (including taxes, insurance, and fringe benefits), vehicle cost, office expense, utilities, secretarial fees, accounting fees, legal fees, bank charges, etc. To determine how much to use as a bid figure, make a detailed list of these and other non job-related items on a separate worksheet. Then figure out what those general operating expenses would be for each item during the period of one year. Add all of these items up and divide by the number of working days in a year. This will tell you how much the overhead is for one day. Use this figure to determine the amount added to the bid based on the time required to complete the project. Don't let hidden costs get away from you. Otherwise the overhead will not be adequately covered and will eventually come out of the profit.

PROFIT: There are no hard and fast rules dictating what percentage of profit to add. A TIB must use his judgment and knowledge of current market conditions. In a good economy the profit percentage can increase. In a bad economy, when jobs are scarce, the profit percentage may have to be lowered to be competitive. It's not a matter of what the TIB wants for a profit, but how much he can get.

The profit figure is usually 5 to 15% above actual cost. The average percentage for overhead and profit combined is about 10 to 25%. On small jobs, the percentage of profit should be larger. The smaller the job, the harder it is to make a good profit. Certain types of jobs such as remodel work contain an element of risk, and TIBs should add a higher percentage to this type of work.

Every TIB is in business to make money, and EVERY job should produce its share to reimburse him for his services and to provide working capital for the business.

On the other hand, don't overprice the jobs either. Make a fair profit for an honest day's work.

COST PLUS BIDDING

Another form of bidding is called Cost Plus. This means that the client will pay you for the actual cost of the labor, equipment, and materials on the job, plus a percentage for profit added to the actual costs. This method is especially useful on jobs where the extent of the work cannot be determined accurately. This is a common type of bidding in remodel work. For example, suppose a TIB contracts a small remodel job, and there is a cost plus 15% agreement with the owner. During one billing period of a cost plus job there is an accumulated total gross labor cost of $2,105.00 (gross labor cost should include taxes, insurance, fringe benefits, etc.), a total material cost of $1,526.00, and a total equipment cost of $100.00 (owned and/or rented equipment directly attributable to the job), bringing the total cost of the job to $3,731.00. Just add 15% to this which will be a total due you of $4,290.65 as shown on the Job Expense Report (Figure 7). Remember, this 15%, or whatever percentage is needed, must include your overhead; the cost of operating the business. Otherwise what you think is profit will be eaten up by the overhead expenses.

There are a few things to remember when making this kind of agreement. The client will want to spend as little as money as possible, and he may ask you which type of agreement is better for him — the fixed price or the cost plus. Your ability to answer this question diplomatically may determine whether you will get the job or not. Both ways are acceptable, but the TIB has much less risk on the cost plus job. However, if the expenses go beyond what the client had originally anticipated, he may become very upset. On a cost plus job there is a moral obligation to the client to watch how the money is spent as if it were your own. Prove to him that you are protecting his interest.

Keep Accurate Records

It is very important to keep good records for all material, labor, and equipment costs for a cost plus job. Using the

JOB COST RECORD

PROJECT ___Jones Building___

DATE	DESCRIPTION	MATERIAL	LABOR	EQUIPMENT	MISC.
xxxx	City Steel Company	325.00			
xxxx	Williams Hardware	101.50			
xxxx	Frank Jones		252.00		
xxxx	Bill Davis		50.00		
xxxx	Sam Bonnett		310.00		
xxxx	Central Equipment Rental			100.00	
xxxx	Bonded Printing				56.00
xxxx	Frank Jones		275.00		
xxxx	Bill Davis		327.00		
xxxx	Sam Bonnett		298.00		
xxxx	Harvest Supply	299.50			
xxxx	Wilson Door Company	156.00			
xxxx	Tallon Paint	15.30			
xxxx	White Lumber Company	355.70			
xxxx	Frank Jones		212.00		
xxxx	Bill Davis		86.00		
xxxx	Sam Bonnett		295.00		
xxxx	Harvest Supply	120.50			
xxxx	Williams Hardware	92.80			
xxxx	White Lumber Company	59.70			
	TOTALS	1526.00	2105.00	100.00	56.00

PERCENTAGE OF COMPLETION.. 36 %

TOTAL ESTIMATED COST (including changes) .... $ 10,596.00

ACTUAL EXPENSES TO DATE... $ 3787.00

VARIANCE .... $ 6809.00

FIGURE 24

Expense Report (Figure 7), you can see how a history of expenses can be kept for a specific job. As expenses (and client payments) are accumulated, they are shown on this form. An Expense Report form should go to the client with each Invoice (Figure 25). A copy of every receipt (materials purchased and equipment rented) should accompany the Expense Report and Invoice. If you or one of the employees loses a receipt or forgets to properly identify an expense, you may not be able to bill the client for that expense.

In other words, let's say you have three jobs. One is cost plus and the other two are fixed price. If an employee picks up material for all three jobs and the receipt does not show the breakdown of what material went to which job, later, as the bills are received and paid, that load of material may be wrongly identified. As a result, the two fixed price jobs show a higher material cost because it included the material for the cost plus job. The cost plus job, on the other hand, won't show the expense at all. Therefore, the TIB never gets paid for that material. Money is out of your pocket because of poor record keeping.

Establish Labor and Equipment Charges

Before making a cost plus agreement with the client it is important to preset the pay scale for the workers who will be on the job (hourly wages plus taxes, insurance, and fringe benefits). In other words spell out how much you will be charging per man hour. You may want to quote two or three different hourly prices based on the level of workers such as supervisor, craftsman, and helper.

In the same manner quote a daily charge for owned equipment that will be used on the job. This won't be a visible job expense, but it is a cost that should be charged to the client. On the other hand, rental equipment should be billed to the client at cost. These established prices should be written into the contract and/or proposal so that the client cannot dispute the invoice price structure at the time of billing. Refer to the section on contracts for more information.

Time Cards

When keeping track of employees' time on a cost plus job, have each employee use a Time Card (Figure 23) so that the total man hours on each job can be captured. On the weekly time card the employees identify the job name and hours worked for each job. The TIB can use this information when filling out the Job Expense Report. The client may challenge the figures, so be prepared to give him copies of the employee time cards showing the hours worked on his job.

UNIT PRICE BIDDING

For many trades, particularly in construction, it is common to bid jobs by using unit prices. In other words, the TIB will give the client a set price per unit for completing a particular task, such as so many dollars per sheet or per square foot. The price may or may not include the material, depending on the client's desires and the TIB's resources and policies.

The TIB may or may not furnish the material, or possibly he will furnish just the miscellaneous material such as nails, mortar, grout, etc. For example, if you are a framing contractor, you might be expected to price a job with a unit price of so much per square foot for all under the roof area which would include labor and nails only, with the client supplying the lumber. This may be similar if you are a tile setter, mason, painter, etc. If you are a drywall hanger, you might be expected to bid the work at so much per sheet. Many times you may need to establish two prices—one with material included and one without material. Become familiar with the going price in your trade so that you don't overprice yourself out of the market.

When establishing the unit price a TIB must carefully evaluate the actual cost and add enough for a reasonable profit. Be sure to include all of the small incidental costs when establishing the unit price. Evaluate the total bottom line price for the job to cross check the profit margin and make sure the unit price has been established accurately.

Advantages of Unit Price Bidding

Bidding unit price provides some real advantages for certain trades. Unit price bidding makes it easy for the TIB to calculate the costs. The total bid for the job will go up or down based on the number of units on the job. This relieves the TIB from much of the time-consuming process of detailed estimates. It also makes it much more difficult for the client to erode the TIB's price structure. The client has many opportunities to critically analyze the fixed price or cost plus bid, but he will have little to say about the unit price.

Another advantage is giving the client the ability to know the price without a formal bid, particularly if a TIB has completed similar jobs for that client. Unit price makes it much easier for the client to feel as though he has control over his pricing. He may use a TIB on job after job without worrying about the total bid amount. Unit pricing tends to eliminate time-consuming bidding wars.

Disadvantages of Unit Pricing

While there are advantages to unit pricing, there are also a few pitfalls. Unit price bidding tends to distance the TIB from the ability to accurately compare actual costs on previous jobs when bidding new work. As a result, when variations or difficult work require special considerations the TIB may not know how much to adjust the price, higher or lower, depending on the variations. It is often easy to overprice simple jobs and underprice complex jobs.

Client Considerations

Whether fixed price, cost plus, or unit price, preparation of prices and proposals needs careful handling. This is your security for future success. It is not uncommon to find yourself in a situation where a client wants you to give him a price "right now" for work he wants done. If you are uncomfortable with the situation, even if it is a small job, don't feel intimidated into giving the client a price that you are not sure about. It's a common practice to tell the client that you will call him back in an hour, day, or week with a price. Intimidation and pressure can cause overwhelming errors. Be diplomatic with your clients, assuring them of the need for accuracy and proper pricing procedures.

One form of intimidation is when a client wants to stand over you while you are working on a bid. Or possibly he wants to review it with you to see how you calculated the prices. Do not allow your client to watch you bid a job, and do not allow him to review or even see your bid sheet. This is not a common practice, and it is certainly not a smart thing to do. Once they get inside your system, they will pick it apart. Don't let it happen.

These types of situations will eventually come up, so be prepared with your response. Establishing your business policies in advance will eliminate these awkward situations.

Trying to Save a Buck at Your Expense

Sooner or later, you will have a client who wants to help with the work by taking over certain areas of the job to save himself some money. For instance, suppose you bid a job that provided material to complete the entire job. Then the client decides the buy the lumber or drywall, or wants to install the insulation. This procedure is acceptable only if it will easily work for personally. The client should not set your policy, only you can. However, should you decide to split the responsibility, make sure that you review the bid carefully. Your written bid to the client should spell out, in detail, the items that are included by you and the items that are not included. It should also specifically state which items the client is responsible for. Refer to the chapter on contracts for more information on detailing your bid.

The client won't understand how much more difficult it becomes to run a job with his intimate involvement, especially if that client is not in the construction industry. In his eyes, you have less responsibility and less material cost. In reality, you may have certain savings as a result of his purchasing materials or performing a portion of the work, but for every dollar saved you could spend a dollar trying to control the client's activities on the job.

Many clients are not used to flowing with the schedule of a construction job which has strict time and procedural demands. In other words, they may not understand the need for materials to be delivered on time and for work to start and finish on certain pre-scheduled days. Explain to the client, in detail, his responsibility and how he is expected to fulfill that responsibility. He will appreciate your taking the lead. Treat him with the respect he deserves as the client, but put the same demands on him that you would any worker, supplier, or subcontractor.

BIDDING GUIDELINES

Whether a TIB's business is young or old, it's important to know not only how to bid a job, but also which jobs should not be bid at all. Assuming that you can do any job that comes along can have disastrous results. It's easy to get in over your head. Know your limitations and become aware of the hazards that turn simple jobs into catastrophes.

Experience is a tough teacher, and most experienced TIBs will tell you about at least one or two jobs they regretted having taken. Special circumstances and situations must be taken into consideration when bidding each job. Simply put, there are items that affect the job that never show up on the plans and specifications. These special considerations may be serious enough to warrant a TIB's refusing to bid the job. There are some jobs that the TIB should not get involved with at all.

Here are some tips that will help to shape the TIB's bid and make the decision of whether or not to submit a bid.

1. Proper supervision must be provided on every job. Don't get involved unless you can control the work. If it is a large job, a full-time job foreman may be needed, based on the TIB's availability. Make this determination wisely and adjust the bid accordingly.
2. Distance from the home office to the job should always be a consideration. If the distance is too great, there will probably be an additional travel expense paid to the workers. Sometimes the travel time cuts down on the time on the job. Allow for extra days to complete the work if this is the case.
3. Danger is a real and serious factor in construction. If the level of risk to the workers is too great,

do not take the job. If just one worker gets seriously injured, it will make the financial profit seem insignificant. Make sure your bid includes a provision for the proper equipment necessary to provide a safe and trouble-free work area.
4. If there are several bidders on a job, don't be intimidated into lowering your price. Many TIBs feel compelled to lower the price just because someone else turned in a lower bid. Don't worry about losing some jobs to lower bidders because...THERE WILL ALWAYS BE SOMEONE WHO WILL BID THE WORK CHEAPER. This does not mean they know what they are doing or that they will make a profit. Never assume that either is true about any other TIB.
5. Material availability, delivery schedules, and equipment availability should be considered in the bidding process. Be careful not to assume the cost factors and availability of equipment. Always double check this information.
6. If you have plenty of work, don't take jobs that have little profit. Stand firm with the bid that will provide a fair but good profit. When work is slow or when the economy takes a downturn, then you may need to reduce the profit, but not before.
7. Don't go beyond your ability to finance your portion of the work. Find out not only what the client's payment policy is, but also his actual payment track record. You must be able to finance your cost of the material and labor until the client issues payment. If the job is too large for you to handle, don't risk your money or reputation by overextending yourself.
8. If the work is difficult beyond what you are comfortable with, possibly you should wait until you gain the necessary experience. This is not to urge you to back away from a challenge, but everyone has his limits.
9. Try to avoid jobs that will draw you away from work that you otherwise would make more money doing, or jobs that keep you from doing work for a reliable and regular customer who will provide more work in the future.
10. Stay away from dishonest clients. If you question his ethics, follow your instincts and back

away. Watch out for big talkers and silver-tongued messengers of "hype." Promises of riches and easy payments usually will vanish. Don't get caught holding the bag. If you are unsure of the client, you should ask for the money up front, or preferably turn the job down. Being fooled by the mood of the moment is certainly no way to make decisions. Look at his track record and assess the situation wisely.

11. A major factor in determining labor costs and scheduling is the season of the year through which the job will run. It is necessary to evaluate the number of days required to complete the work, based on projected weather conditions.

Site Conditions

Depending on the trade involved and the type of job there are special site-related details that must be considered on every bid. Whether the bid is fixed price, cost plus, or unit price, these special considerations are variables that cannot be specifically identified on the bid as cost items, but they will directly affect the TIB's costs as the job progresses.

For many trades such as excavation, landscaping, framing, utility work, and concrete work the site conditions are of great importance. Visiting the site is essential if the TIB's trade is directly affected by its conditions. While at the site, all existing conditions must be examined carefully to evaluate how the bid price is to be established. For example, the following conditions should be observed:

1. General site conditions — Look for existing structures that must be moved, access to the job from main roads, conditions of existing curbs and streets, obstructions such as signs, trees, and electrical poles. The soil conditions may need to be checked for consistency.
2. Location of the site — Check for accessibility to the site by material trucks. If the distance is a problem, will workers be available? Will subs charge extra to drive that distance?
3. Utility locations — Find the nearest water and electrical power supply. Are restroom facilities available on the jobsite?
4. Material storage — There must be a place to store

the materials during the course of the job. Will the trucks be able to get to the storage area, or will the materials be moved by hand?
5. Trash — Find the nearest dump location.
6. Safety — The neighborhood conditions may require extra precautions to safeguard the job. Equipment and material may be stolen unless properly protected.

Retainage

Special conditions must be taken on all jobs that withhold retainage. Retainage is a percentage of what a client owes the TIB but holds back until a predetermined time. Let's say a TIB has a contract agreement that specifies a 10% retainage. This means that on each progress payment the client will hold back 10% of payment until thirty days after the job is complete or whenever the contract specifies. If the TIB bills the client $1,500.00, the client will pay the TIB only $1,350.00. The remaining $150.00 (10%) will be held by the client until a later time as specified by the contract.

As you can see, the client can be holding a significant amount of money for what may be a long period of time. The percentage will vary, as will the scheduled date of release of the retainage money. Obviously, if the contract amount is large the TIB may not be able to afford the retainage agreement. As a result, the TIB may need to include extra money in the bid to allow for the delayed payment. Refer to the Contracts section for more information on retainage.

TAKING BIDS FROM SUBCONTRACTORS AND SUPPLIERS

As part of the bid process, the TIB will request and receive bids and price quotes from subcontractors and/or material suppliers. Keeping these bids and prices organized and recorded will help eliminate such future problems as overbilling by subs and disagreements over what price was quoted.

As bids and price quotes are received they can be listed on the Subcontractor/Supplier Bid Sheet (Figure 8). This form will keep an accurate record of all bids and prices for a particular job. Keep a separate bid sheet for each job so you can easily compare prices. When the TIB is satisfied that he has received the best available quotes, he can transfer the selected prices to the Bid Worksheet (Figure 6). After entering the numbers on the Bid Worksheet, check the appropriate space on the Subcontractor/Supplier Bid Sheet showing which sub or supplier was selected. It's easy to forget which price or company was selected two months later, so making note of it will prevent mistakes.

This form should also be used in conjunction with the Purchasing Worksheet (Figure 2). Refer to the Purchasing section for more information.

Always Get a Written Quote
From most subcontractors and some material suppliers it is important to get a written quote. In analyzing the sub's bid, the first thing to do is to determine whether the bid is complete. It's common for subcontractors to submit a price that is incomplete, covering only a portion of the work. Sometimes it is because they don't understand the scope of the work, or they didn't take the necessary time to search out the details in the plans and specifications.

Another reason to get a written quote is to verify whether the bid includes labor, materials, or both. Assuming the sub understands his responsibilities without the TIB's careful evaluation can be disastrous. For example, the TIB may use a bid for tile setting in establishing his price for the client only to find out later that the bid included labor only. Written quotes will eliminate this problem. Normally the suppliers will alert the TIB of future price increases and will hold a price quote for a certain length of time. Write the guaranteed price time limit on the Bid Sheet so that you will know how long to hold the client's bid price. Usually they will give a thirty to sixty day price guarantee.

Telephone Bids
Very often last minute subcontractor bids are received over the telephone. Obviously there isn't enough time for the sub to get a written quote to you. An efficient way to record these bids is by using the Telephone Bid Sheet (Figure 9). By recording the bid in this manner mistakes can be avoided. Fill in the Telephone Bid Sheet just as the sub or supplier would fill out his written quote. Ask him to follow up with a written bid.

SUBMITTING THE BID

The manner in which a bid is submitted to the client can be as much a determining factor for the client as the price. He will be looking for both knowledge and individual character in each bidder. Therefore, your attitude is extremely important. Arrogance, rudeness, and hype are obvious turnoffs. On the other hand, so is lack of confidence and insecurity. Try to be relaxed, courteous, responsive, alert, and prepared.

When negotiating with the client, certain things should not be said such as, "We normally don't do this kind of work, but we will bid the job anyway." What this says to the client is, "If you are a big enough fool, you can hire us to do the work." Another statement to avoid: "We are so busy right now, it's hard to keep up with everything." What this says to the client is, "I'm too busy to do this job, and I won't be able to give it my full attention." Also: "Things have been so slow I've had to let most of my employees go." This says to the client, "If you hire me to do the work, I may not have experienced employees on the job." One more to avoid: "I really need the money in a hurry." This says to client,"If you hire me there may be liens filed against your project, because I may not be able to pay the material suppliers due to my desperate financial situation."

Most of these statements may give the client the indication that talking with you is a waste of time. Certainly a TIB should never lie about the capabilities of his company, but if you have decided that bidding the job is right for you, then your job is to sell the client on why he should hire you.

Cover Letter
An excellent sales tool is a cover letter (Figure 10) attached to the front of the proposal/bid. It should be very simple — stating your appreciation for the op-

portunity to bid the job. This is a very effective way of letting the client know you care about him and want to work for him. You should also included a list of past jobs with client names and phone numbers to show the client your track record. Whether he calls any of the references or not, he will be impressed that you have a history of similar work and that you present yourself so professionally.

5. CONTRACTS AND PROPOSALS

WRITTEN DOCUMENTS

There are different types of contractual agreements, each serving a different purpose. The way a contract is written depends on the type of work and the kind of relationship between the TIB and the client. Certainly, an effective way for a TIB to present himself professionally and develop credibility is by his ability to understand and write contracts and proposals. Contractual agreements don't have to be lengthy and complicated, but they most certainly are more than writing the price on a napkin or giving it verbally. You would be surprised how many TIBs operate in this way. This opens the door for several serious problems to arise.

Without a written contact or proposal, the client can easily misunderstand or assume that a much larger scope of work is included in the price. For instance, if the TIB gives a verbal bid of $2,500.00 for plumbing work, the client may assume the bid includes fixtures (such as sinks, toilets, and water heaters) when in fact it only included labor to install the fixtures. When the client orders work to be done, the TIB may be accused of misleading or trying to cheat the client after he finds out the fixtures cost extra.

Even worse, if the TIB completes the work and, after buying the fixtures, sends the client a bill for $2,500.00 under the original agreement and an extra charge of $1,500.00 for the fixtures, the client will hit the ceiling when he sees the bill. Because he assumed the fixtures were included in the original bid, he may even refuse the pay the extra, claiming the TIB overcharged him.

In a similar situation, a TIB verbally bids $3,200.00 for excavation work. After the job is completed, the TIB submits the bill. When the client inspects the work, he claims that the original price included several truckloads of sand. The TIB's price did not include the sand, and if forced to provide it he would lose money on the job. The client insists that he get the sand and refuses to pay the bill until the job is complete. Without a written contract the TIB has no legal rights.

Another problem with verbal bids is that the client may forget what price was quoted. After the work is complete and the bill is submitted, the client claims that you quoted $1,585.00 and not the $1,875.00 shown on the invoice. It's his word against yours. Without a written document signed and dated by both yourself and the client, you have little or no legal claim to payment for work done.

Without a written, preferably typed, proposal, you may not be accepted as a quality contractor/businessman. It may seem absurd that a piece of paper will cause the client to think that you do good work, but the TIB who pays attention to details is the tradesman the clients will hire. This holds true for every aspect of your business.

There are pre-printed proposal and contract forms

available from several suppliers. Refer to the Resource Materials section for a list of these suppliers.

Never Start Without a Written Document

Be sure to never, ever, ever start work on a job without a signed original of the proposal in your files. Even if a proposal has been given to the TIB it is not a legally binding document until he signs it, and without this document in your hands you are asking for big trouble by starting the job. If a TIB starts the job without the contract in hand, his guarantee of being paid for the work is limited. Never assume the client will always do what he says. Just because he says to start the work and he will put the proposal in the mail doesn't mean it will happen. It would be nice to trust everyone, but unfortunately you can't.

The signed proposal/contract is the only document that legally proves that the client agreed to the price and gave you instructions to proceed with your portion of the work.

ESTIMATES

The term "estimate" is used in two ways. An estimate is the bid worksheet; what the TIB uses to formulate the bid price. The estimate, as it relates to contractual agreements, refers to a preliminary price given to the client before a firm price can be established.

An Estimate (Figure 11), even though it is submitted to the client, should not be treated as a legally binding agreement. It simply states what work will be done for what price. Estimates are typically given in the preliminary stages of a project or before you have access to all the information needed to formulate a proper price.

Many times, before the architectural plans are complete an estimate is given so that the client can get a feel for what it would take to complete the work. When giving a written estimate, do not use a formal proposal form. Do not give any indication that this is a binding agreement or a final contract price for the work. Use regular typing paper or letterhead, and send the information in the form of a letter. Write it very clearly in the letter or estimate, "This is a preliminary price and is subject to change. This is only an estimate."

Do not sign the estimate making it a legally binding document. Because an estimate is sometimes given before the drawings and specifications have been completed, you must be very clear as to what criteria you used to establish the price. In other words, call out certain specifications on equipment and lay out the scope of work in as much detail as possible in a letter format.

If you need further information to be sent with the estimate, write a separate letter, sign it, and attach it to the front of the estimate. Make it very clear that this is not a binding agreement. Again, when giving a preliminary price to a client, be very careful that he understands on what information your estimate is based. He may be upset if the final contract price comes in much higher than the estimate. You may lose the client if this happens, even though you can justify the price increase.

You are not obligated to give these preliminary prices, but many times it will be necessary in order to please a regular client or to help build a relationship with a new client. The client needs as much help as he can get, and your ability to formulate a professional and detailed estimate will many times firm the client/TIB bond.

PROPOSALS

A proposal, if signed by both the client and the TIB, is a contract which is a legally binding agreement between two individuals and/or companies. The Proposal Form (Figure 12) is the type of agreement that a TIB can regularly use, particularly in the early years of his business. This proposal form can be found in most office supply stores. It will have pre-printed information that is critically important and might otherwise be overlooked, such as length of time the proposal is valid, warranty, and payment terms.

Let's explain the different parts of a proposal step by step.

What Is Included in the Price

You are proposing to furnish material and labor to complete certain work for a certain price. Be specific

Proposal

Page No. 1 of 1 Pages

PROPOSAL SUBMITTED TO J & K Construction	PHONE 777-9999	DATE XXXX
STREET 94 East Street	JOB NAME BFK Building	
CITY, STATE AND ZIP CODE Your Town, USA xxxxx	JOB LOCATION HWY 93	
ARCHITECT Goldsmith & Associates DATE OF PLANS 6/1/86		JOB PHONE

We hereby submit specifications and estimates for:

Finish out office space according to plans and specifications by Alexander & Company dated 6/1/86. Work in office area will include:

Wood framed walls w/5/8" fire code sheetrock. Exterior walls to have R-19 fiberglass insulation. Furnish and install interior doors, as specified. Acoustic ceilings, cabinet in snack room, and wood base and trim - as specified.

This proposal does not include:

Tape, bed, texture & paint walls, doors or trim.
Restroom hardware, plumbing, and fixtures, electrical and fixtures.
Permits and fees.

The undersigned guarantees work and material finished by us against defective workmanship and materials for a period of one (1) year from the date of acceptance. Defective work becoming evident during the period of guarantee, upon written notice of owner, will be promptly repaired at no expense to the owner.

We Propose hereby to furnish material and labor — complete in accordance with above specifications, for the sum of:

Six Thousand nine hundred and fifty and no/100-------- dollars ($ 6950.00).

Payment to be made as follows:
Billing is done by the first of every month for work completed during the previous month. Payment is due within ten days after billing.

All material is guaranteed to be as specified. All work to be completed in a workmanlike manner according to standard practices. Any alteration or deviation from above specifications involving extra costs will be executed only upon written orders, and will become an extra charge over and above the estimate. All agreements contingent upon strikes, accidents or delays beyond our control. Owner to carry fire, tornado and other necessary insurance. Our workers are fully covered by Workmen's Compensation Insurance.

Authorized
Signature _____

Note: This proposal may be withdrawn by us if not accepted within _____ days.

Acceptance of Proposal — The above prices, specifications and conditions are satisfactory and are hereby accepted. You are authorized to do the work as specified. Payment will be made as outlined above.

Date of Acceptance: _____

Signature _____

Signature _____

about what items are included. Don't let details fall through the cracks. If only certain labor and certain materials are included in the price, then list what is included, so that the client has a full understanding. Do not rely on what was said during a discussion with the client, but rely solely on what is written in the proposal. Identify which documents were used to arrive at the price, such "Plans and Specifications by Alexander & Company, dated June 1, 1987."

Many times the client will request that you break out certain items on the bid and list a separate price for that item. First show the base bid and then list any adds or options and show a price for each item. If the client chooses to include this item in the price, simply add the extra work to the base bid when completing the billing procedures.

What Is Not Included in the Price

It is as important to list what is not included as it is to list what is included. Too many times the client will assume that the proposed price includes more than you intended, not because of what is written, but because of what is not written. Clients have a bad habit of expecting something for nothing. Therefore, list in detail which items are not included that relate to your particular trade. Then list any items that are not directly related to your trade, but are affected by your work. This will eliminate any misunderstandings.

Warranty

All work should have a warranty. In today's market, too many companies do not stand behind their work. Be a businessman who cares about his work and his client. The day of the "fly-by-night" contractor is over. It's time for "service after the sale" to mean something. It is a standard practice for the TIB to correct defective work or work not conforming to the plans and specifications (identified in the proposal) for a period of one year.

Many times if your work involves appliances, fixtures, and equipment, the manufacturer will provide a separate warranty for these items. If an air conditioning unit goes out during the twelve-month warranty period you will be responsible for having it repaired. However, the manufacturer and dealer will usually pay for such repair. At times like this, you will be glad you have developed a relationship with a reputable dealer who stands behind his products and who has an in-house repair department.

Warranty work can become a serious problem, particularly if the work is not done properly the first time. That's why it is important for all the TIB's suppliers and subcontractors to supply him with a similar warranty. If the subcontractors don't warranty their work, the TIB could have some overwhelming expenses repairing a bad job.

Specifying that the TIB's work carries a written, limited warranty serves another very important purpose. It limits your guarantee to defective workmanship and materials, but does not guarantee damage from abuse, misuse, tornado, etc. Wording might be, "The undersigned guarantees work and material furnished by us against defective workmanship and materials for a period of time of one (1) year from date of acceptance. Defective work becoming evident during the period of guarantee, upon written notice by Owner, will be promptly repaired at no expense to Owner." A written limited warranty is much better than an implied warranty. In other words, if the warranty limits are not specified, the TIB may become responsible for a much broader area of warranty obligation.

Payment Schedule

In your proposal, it is necessary to spell out in detail what is expected regarding the scheduled payment for the completed work. Here are a few standards methods of payment. Use the one that best fits your company and situation.

1. 50% halfway through the job and 50% at completion.
2. 20% down payment before starting the job, with scheduled progress payments throughout the job.
3. Billing is done by the first of every month for work completed during the previous month. Payment due within ___ days after completion of the work.
4. 100% due upon completion of the work.

Acceptance of Proposal

On all proposals it is very important to state how many days from that date of the proposal the bid price must be accepted by the client before it becomes invalid. In other words, material prices, workloads, and economic conditions are continually changing. Your bid is based on certain specific information. This information can change drastically within a few weeks or months, and if the prices go up, you cannot be bound to a contract that was figured using an outdated price structure. Most commonly, proposals are good for ten to sixty days.

Another reason for setting a time limit on the proposal is that it causes the client to make a decision. If he knows that the price may change, he will be more interested in signing the proposal before the time runs out. If the time does run out and the client wants to sign, be sure to double check current material prices to see if the proposal price is still good.

CONTRACTS

As stated previously, a proposal (Figure 12) can be the contract between you and your client, and is acceptable since the format is most frequently used, especially in the early years of a TIB's business. However, from time to time the client will require a contract be used that is several pages long and full of legal wording. Your first reaction may be to pretend to understand what it says and sign it, knowing full well that you don't understand it. You shouldn't be embarrassed if you don't understand a contract, and you certainly don't want to sign anything that you don't understand.

If you take the time to read carefully, most contracts can be understood because they usually contain fairly standard information. By becoming familiar with the basic topics you can learn to read through a long contract with full comprehension. Here are some topics or categories you might find in a contract:

Scope of Work

The scope of work is a list of the work and actions of the TIB that will be performed on the job. It spells out his responsibility including items such as city permits, conforming to local or state building codes, etc. The scope of work also identifies specific items for which the TIB will not be responsible.

Contract Amount and Terms of Payment

This section states the agreed price for the work to be done. It can be shown as a fixed contract price or as a unit price (such as $1.50 per square foot). The payment schedule is usually addressed in this section. It will list how much, and at which stages of the job, payments will be issued. Be very careful to note exactly what the contract says about payments and possible retainage of a percentage of each payment.

Retainage

Retainage, normally addressed with the Terms of Payment requirements, can have an effect on how the TIB formulates his added profit. Retainage is exactly what it sounds like: it is money retained. For instance, if the contract calls for a 10% retainage until thirty days after the work is completed, it means that as invoices are sent to the client, he will hold back 10% of everything due the TIB. The 10% retainage will be paid after all of the TIB's work is completed and accepted.

In some instances, especially on larger projects, a retainage may be withheld until the total project is completed and accepted. This can be a lengthy period of time, especially if the TIB's portion of the work is early in the progress of the job. If the total job is delayed, the TIB may wait months before he receives the final payment. The way the contract is worded can be difficult to understand, but a TIB must be sure of when the retainage will finally be paid.

The amount of retainage and the scheduled date of payment will vary from contract to contract. Since the client is holding so much of the TIB's money, he must make sure that either the bid is padded to make up for the delayed payment, or that he can financially withstand the extended period of time before the final payment.

Change Orders

A change order is a document issued by either the

TIB or the client (Figure 13). It identifies a change in the job that financially changes the original agreement. Suppose the client wanted to add a window in an outside wall. The change order would be written listing the scope of work and the cost of the change. The client should sign each change order so as to give written approval to complete the additional work.

Change orders are considered to be an addendum to the original contract. Therefore, the original contract price will decrease or increase based on the change order.

If additional work is added to a contract that specifies a definite time schedule, then the change order should also include an extension of time for the completion of the extra work. If special material has to be ordered or fabricated, or if the building has to be expanded, the delay could be substantial.

Writing change orders can be time consuming, but they are very valuable documents that will keep you out of hot water. There are very few jobs where the client doesn't make some kind of change that results in an increase in the cost of the job. If you use a change order each time a change is made, it will help you keep track of the billing, and it will also prevent the client from becoming overwhelmed by a bulk of extra charges at the end of the job.

Delay
Read this section of the contract carefully. Delays can put you in jeopardy, particularly if they are caused on the job by you or your employees, resulting in a violation of the contract agreement. Most of the time you're dealing with lost time because of weather or omissions by the client. However, for your safety you should keep a daily log for every job. In the log list the daily weather conditions and how they affected the job. That record can save you much anguish if the job goes past the deadline date by verifying that weather was the cause for the delay.

Don't count on the client to remember how the weather delayed the job. All he knows is that you are not finished. The daily log will be a record that will determine whether or not you violated the contract. You can also use this log to make a record of how the

client's changes affected the schedule of the job. Remember, written records are extremely important. Written documents can mean the difference between a satisfied customer and a court date.

Hold Harmless Agreement (Indemnification)
This agreement allows the client to make sure that he or any of his employees are not held responsible for any damage to property or personal injury that might arise in connection with or as a result of your portion of the work. It may also state that you cannot legally start work on the project until you have furnished the necessary certificates of insurance required.

Governmental or Code Compliance
The work that any TIB does on a job must comply with all applicable state, county, and city building codes. You must also comply with all governmental regulations concerning workmen's compensation, payroll taxes, unemployment insurance, Social Security, etc. This section is usually a catchall that basically means if you do this job, you must comply with every law that encompasses your work. This is so the client won't be held responsible if you, in some way, do not comply with the regulations.

Performance and Payment Bonds
This section spells out the possible requirement for a performance or payment bond. These bonds will guarantee the client that your portion of the job will be completed properly and that all of your bills for that job will be paid. See the section on Bonding for more details.

Insurance
The TIB is responsible for providing the proper insurance for his company. In most contracts, the client will list the kinds of insurance required in order to work on his job. Normally, only three types are listed: General Liability, Auto Liability, and Worker's Compensation. The limits listed in the contract won't necessarily be the recommended limits for your company, but only the minimum required for this contract. Visit with your insurance agent to determine the right coverage for your company. Refer to the chapter on Insurance for more information.

CHANGE ORDER

RF BUILDERS   1000 Rosewood   Your Town, USA

(OWNER): J & K Construction	PROJECT: BFK Building	CHANGE ORDER NO: 1
ATTENTION: J. Barnes	LOCATION: HWY 93	
		DATE: xxxx

This change order is hereby incorporated into the original contract and is to be attached thereto. All other items and conditions of the original contract and any prior change orders not modified below remain the same. Changes are as follows:

ITEM #	DESCRIPTION OF CHANGE	ADDITIONS	DEDUCTIONS
1	Added special entry door (model #100 Wilson Co.)	328.00	
2	Raised ceiling height from 8' to 9' in rooms 3 & 4	375.00	
3	Omitted window in bathroom		65.00
	TOTALS	703.00	65.00
	NET CHANGE BY THIS CHANGE ORDER	+638.00	

Original Contract Sum ....$ 6950.00

Net Change by Previous Change Orders ....$ -0-

Net Change by This Change Order ....$ 638.00

Revised Contract Sum...$ 7588.00

This Change Order Increases the date of substantial completion of this job by _____two_____ days.

Accepted By: _____

Contractor

_____

Owner

FIGURE 13

Warranty

Much about warranty has already been covered. However, in a contract you should verify what your warranty responsibility is. Don't take this lightly. You could finish a job and four years later get a call to come fix a door that is not shutting properly, just because your contract stated that you provided a five-year warranty.

Most warranties are for one year against defective materials and workmanship. As a reputable TIB you may see the need for affirmative action on problems that are beyond the scope of the stated warranty. This will prove to the client your commitment to customer satisfaction.

General

Every contract has its share of legal talk that won't necessarily directly affect you. There will be information about where papers are to be filed, the termination of the agreement, and that all work will be done under the laws of the state, etc. In most cases, this information is the same from contract to contract, but if you don't understand it, don't sign until you do.

Notary

Every contract is signed by the TIB and the client. These signatures are usually witnessed by a notary who will put his or her official seal on the document, making it legal and binding. Notaries can be found with little effort. Many companies have at least one secretary who is a notary. Banks and insurance agencies usually have notaries on their staff who will be glad to help you for a small fee.

Negotiating

It is important to note that just because a client has prepared a typed contract ready for your signature does not mean that it is worded properly. You can change anything on the contract that you feel is not in your best interest. This is called negotiating. The client may accept the change, or he may accept only a portion of it. This give and take negotiating may take time in order to get the wording correct, but it will be time well spent.

If In Doubt, Seek Professional Advice

Not all of the many different items that can be found in contracts have been discussed here, but with this basic understanding of the format of a standard contract, you should be able to negotiate with the client on his level. You don't have to be a lawyer, so don't try. If you need help to understand a contract, take it to your attorney. A few bucks spent in his office might save you thousands of dollars in the long run.

6. FINANCING

BORROWING MONEY

The very best way to fund or start up a small business is with your own money. By using your own money, you don't have to pay interest on a bank note. If you don't think interest is a problem, just talk to someone whose business went through an economic downturn and who had a bank note with interest due on a regular basis; see if it was a problem for them.

In a good economy, bankers will loan money to just about anyone. Just because a bank will lend you money doesn't mean it's right for you. Remember, you will always have to pay it back, and you may have to pay it back in a down economy, when sales are low and money is tight. When the banker becomes your financier, he also becomes your partner with a say in what you do and how you do it. They want to be sure you will be able to pay back the borrowed money, and they will involve themselves in your business until they feel comfortable about how you run it.

Plan your business growth, not just for the immediate future, but for the next several years. Don't borrow money to buy new equipment, rent a big office, lease a new vehicle, etc. Operate your business on the money you have, not on the money someone else has. In the long run, you will have more control over the direction and success of the business, and you will also have a substantial advantage over your competitors. Their bids may include a much higher overhead factor in order to cover the interest payments on their bank notes.

Having warned you of the evils of borrowing money, let's face the fact that sooner or later most businesses will need to borrow money, possibly to fund a particular project. Loans to fund projects run from six months to one year, depending on the length of the project. Other loans run one to twenty years depending on the reason and type of loan. As a rule, most lenders will not allow the loan to exceed one-third of the capital of the business. In addition, you will have to show that the projected earnings (future sales) will be sufficient to repay the loan within a specified period of time. This will all vary depending on the type of financial institution. There are sources for loans other than banks.

Sources for Loans

Commercial Banks — Banks generally require collateral that is valued at more than the amount of the loan. If the loan is granted, they will retain liens on the collateral which can be such items as inventory, equipment, or real estate. Commercial banks usually have strict terms. They may ask for regularly updated financial statements and will want the TIB's working capital (cash) to remain at specified levels.

Suppliers — Sometimes your suppliers may be willing to work out some form of loan as long as they feel that you are likely to repay the debt. Your staying in business is to the supplier's advantage, so when things get really rough an honest discussion with your supplier might be a way to keep material flowing to your business. The terms will vary and are generally arranged by mutual consent.

Insurance Companies — Another source for borrowing money may be insurance companies, provided that you can offer sufficient collateral. The payment terms of these loans are similar to commercial bank loans, but sometimes the loans can be arranged for longer periods of time.

Sometimes insurance companies and finance companies will buy up the business's fixed assets and lease them back to the business. This type of lease-back loan is helpful to those businesses which have most of their capital tied up in equipment and property.

Small Business Administration — The SBA does not make loans unless the applicant has been turned down by a bank or other conventional institution. They provide both intermediate and long term loans at reasonable rates. If you're unable to secure a bank loan, you can apply directly to the SBA for up to $150,000, or they also can work with your bank by guaranteeing up to 95% of the loan. Therefore, the most the bank would lose would be 5%. Even with this guarantee. some banks are reluctant to make this type of loan as they feel that a business loan under this arrangement is somewhat risky. Nevertheless, this might be something to look into if all else fails. It is important to remember that SBA loans usually take weeks or months to get approved. This might not be the right source if you need the money right away.

For more help and information on other available financing you can write the National Association of Small Business Investment Companies, 512 Washington Building, Washington, DC 20005.

Whatever form of borrowing money you choose, it is very important to have a specific plan. Know exactly how much money you need and the length of time needed to repay the loan.

BANKING TERMS AND DEFINITIONS

Understanding the rules and knowing vital banking terms will enable you to walk into a banker's office and confidently carry on an intelligent conversation on his level. All you need is a little confidence, which will come as you learn the necessary information. Here is a list of a few basic terms and definitions used in the banking industry.

Bank Note — This is a loan the bank makes to you. They will transfer money into your checking account (or the account you specify) after they approve the loan and after you agree to the terms they ask. You will be required to

sign the loan papers, but if you don't understand these papers, don't feel stupid. Just be honest and tell the banker that you need time to read the papers and understand them. Most notes are the same from loan to loan, but occasionally they will toss in a special notation that could cause you trouble. If necessary, get advice from an attorney or another experienced businessman who understands bank notes.

Collateral — When you make a loan (note) at the bank, you are required to put up some collateral to secure the note. Collateral is something of value that you own (such as automobiles, real estate, certificates of deposit, savings, etc.) which the bank will require you to pledge as security for the note, so that if you default (don't or can't pay the note), they will have the right, whenever they want to, to take that collateral and sell it to help pay off the note. If what they get for the collateral is not enough to pay the full amount of the note, they will also have the right (if you sign a personal guaranty, which you probably will) to take you to court to sue you for the deficiency (the difference between the amount of the note with interest and what they got when they sold the collateral).

Just remember, they will get their money if they can. An important rule for putting up collateral: only put up what you are willing or can afford to lose. There are no sure things, so don't put up your property as collateral, unless you (and your family) can afford to lose it. Also, while the bank holds your collateral, you cannot see it or spend it. Whatever you use as collateral must be something that you don't need to liquidate for an extended period of time (until the note is paid off).

Lien—There are several types of liens. First, in the case of collateral, if you pledged some real estate as security for a bank note, the bank's lawyers will file a real estate lien against the property. This means simply, if you want to sell the property, you cannot until the bank files another set of papers releasing the lien and giving you clear ownership of the property after you pay off the note. If you don't make good on the note, the lien is the one thing that gives the bank the right to take your property. Filing a lien means that the bank's attorney takes the lien (which is signed by the bank and by you and is notarized) to the local governing body (usually the county) and makes it part of the public record.

The second and most important type of lien is a mechanic's lien. Basically, this type of lien is filed by a contractor or supplier's attorney when they have completed work or supplied materials or labor on a project, but have not been paid. Refer to the Legal section for more information on mechanic's liens.

Asset — Whatever you own that is of value is called an asset. Assets are real estate, accounts receivable, inventory, insurance policy cash value, automobiles, cash, stocks, bonds, boats, etc. When the bank is determining whether or not they should loan you money, the first thing they look at are your assets. This is to see what you have that will help you pay off your bank note in case you cannot pay if off as originally planned.

Liabilities — If you understand assets, think of liabilities as just the opposite. Liabilities are what you owe other people, such as bank notes, accounts payable, your home mortgage balance, and other financial obligations.

Financial Statement — When you try to borrow money, your banker will ask for a financial statement (Figure 14). This is simply a list of your assets and a list of your liabilities. If you subtract your liabilities from your assets, you can determine your net worth. Consult an accountant for help and direction in preparing a financial statement. The true financial statement consists of a Balance Sheet and Statement of Income. The TIB should consult the accountant if a more detailed and accurate statement is required.

Debit — When the banker says, "We will debit your account for the legal fees," he means they will automatically remove from your account the amount needed to cover certain fees.

Credit — When the banker says, "We will credit your account for the amount of the loan," he means they will automatically put part or all of the loan amount into your account.

Draw — Certain bank notes, especially interim construction notes, are structured so that you will receive the money in scheduled segments instead of all at once. These scheduled payments are called draws. Usually draws are issued to you based on the percentage of work completed on a project. For instance, if you were approved for a $10,000.00 loan for a project and you have completed approximately 40% of the work, then you can make a draw for $4,000.00 or 40% of the original loan amount. A sample Application for Draw is shown in Figure 15.

Draws are very simple once you understand them. Always keep good records of how much you have completed, so you can make the draws accurately and at the appropriate time in order to have funds to pay for the labor and materials on the job. Even if you don't have an interim construction bank note for a job, you may have to issue an Application for Draw directly to the client if he is financing the project. This type of draw is usually used by general contractors or builders. For the typical TIB in the typical contractual agreement, the function of an Invoice (Figure 25) is simpler and more efficient than an Application for Draw. Either can be used, depending on the type of work and the circumstances involved. The invoice is the preferred form to use, as the primary function of the draw application is to obtain bank draws.

These aren't all of the banking terms you will eventually need to know, but by familiarizing yourself with them, you will be able to better understand what your banker is saying. You will also have a much better opportunity to get what you want.

WORKING WITH BANKERS

It is important for a borrower to know the kinds of loans a bank normally offers. For instance, there are banks which specialize in making loans to people who are involved in construction or oil, as opposed to those who might invest in art. Still others give "conventional" loans only. A bank has to feel comfortable with your type of business. They also want to feel comfortable in their ability to sell the kinds of things they take as collateral (in case you default on the note). Banks, therefore, tend to specialize as they gain experience and knowledge of certain industries.

Unfortunately, a bank's experiences may lead it to tighten its grip on the money supply. For instance, if a particular bank has had several construction businesses default on notes, they will be very cautious about lending money to the businesses in that industry. They also like to deal with people who have accounts in their bank. Asking for a loan without an account in that bank may make it more difficult to get.

Plan Ahead, Prepare to Meet Your Banker

When visiting your banker, be prepared. Those who prepare ahead of time will appear much more knowledgeable and professional. Know what you want, and have it spelled out on paper. Show how much you need, what it is for, and when you need it. Have copies of contracts, information on clients, financial statement (Figure 14), business resumé (Figure 16), and the appropriate plans and specifications.

The banker may ask to review your bid worksheet to see how you arrived at the contract amount. If the information is available, as it would be in a public opening of bids, it would serve you well to show him a list of the other bidders and their proposal amounts. This gives the banker information that your bid is not too far below that of your competitors, thus making it a good possibility to be a profitable venture.

Character Qualities That Make the Difference

When meeting with the banker your attitude is extremely important. Don't act as though your being there will make his day. Be honest, courteous, patient, and above all, be on time for your appointment.

It is just as important to respond properly to the banker's secretaries as it is to the banker. If you don't impress them or are rude to them, the banker will hear about it. Don't track mud, don't smoke, don't talk too much, and don't get in their way. Becoming a small business owner doesn't make you better than anyone else. If you don't have a proper attitude, you might as well hang it up right now. You can do all the right things but with the wrong attitude, and your success, if any, will be short lived.

As you meet with the banker, he will have one main question in mind that must be answered. "Does this TIB have the capabilities and resources to pay back the loan?" He simply wants the confidence that you have the character, experience, know-how, and business management skills to follow through with the proposed plan of action. He will be watching you very closely to try to read your personality and character. How you present yourself will be very important. Refer to the Dress and Appearance section for information on proper attire when visiting your banker.

SECTION III
WORKING WITH PEOPLE

7. EMPLOYEES

HIRING

Employees are the backbone of the business, but finding quality individuals can be a difficult task, and as the business grows it is important to develop the ability to know both how and when to hire employees. Increasing the overhead prematurely can eat away the profits, and hiring the wrong individuals can put an unnecessary strain on the business. Here are a few tips to remember.

Hire Only What You Need
First, evaluate the need for extra help. The workload must be heavy enough to justify an additional worker, not the workload you would like to have, but what you have actually had over the previous months along with the workload you actually have at that time. After the need for an additional employee is realized, it will then be necessary to determine whether a skilled worker or a less expensive and inexperienced employee is required. Many times the skilled workers can become much more productive by hiring a lower paid helper to do the unskilled tasks that up to that time have been the responsibility of the skilled worker. Use caution when hiring experienced help. It may bring on a tendency for the TIB to become less productive himself. Don't hire someone to take your place just so you can relax. A successful TIB is a working TIB. The point is to be as economically efficient as possible when hiring employees.

Second, evaluate your company's ability to support a new employee financially by comparing the additional expense of overhead versus the ability to complete more jobs more quickly. Determine if the overhead bid into the jobs will cover the additional wages, taking into consideration that the extra employee will, hopefully, decrease the number of days to complete the job. Make this decision intelligently, considering all the facts.

Pay According to Skills and Market
Become aware of the going rate for laborers, both skilled and unskilled. Underpaying employees will hinder your efforts to get quality people, and overpaying will hinder your efforts to make money. Workers must, however, be paid a fair wage, one that allows them to pay their bills. An employee who makes enough to live on will be a better worker than one who constantly worries about his financial troubles.

Whether you like it or not, you have accepted the awesome responsibility as the only individual providing the income that pays for his family's living expenses. In light of that responsibility, if you can't pay a worker what he is worth in order to adequately feed and care for his family, then don't hire that individual, or at least find a way to increase his wages, possibly by increasing his skill. Being an employer is more than just signing the paychecks and supervising the jobs. You don't interfere in the employees' personal lives, but you do have a certain responsibility for their well being.

Finding the Right Employees
There are several ways to find a good employee. The most desirable way is through a recommendation from a present employee, material supplier, trustworthy competitor, or friend whose opinion you value. However, care must be taken if hiring a close friend of a present employee. Sometimes jealousy can crop up in other employees who may think that you are

playing favorites by hiring the employee's friend. Often, unless it is carefully controlled, too close a relationship between friends may hamper productiveness on the job and create divisions in the workforce.

An advertisement in your local newspaper usually produces several inquiries. Make the ad very clear as to what qualifications are required, so that you can quickly eliminate those who do not qualify. You might indicate definite hours that they should call, so that you can always be available to talk with them. Try to screen each applicant over the phone before you have the face-to-face interview. This will save valuable time by eliminating the bad applicants before having to go through the interview.

In some states you can notify the local office of the Employment Commission to inform them that you are now hiring. This governmental organization will usually screen the applicants and send you ones that they feel meet the requirements you have requested.

Application for Employment
Have each applicant fill out an application for employment (Figure 17). This form will provide the needed information about the applicant which will help the TIB make his decision. Keep these on file so that if the person you hire doesn't work out, you will have the information available and can quickly hire a replacement.

No matter what means you use to locate new employees, after the interview be sure to check the references of each serious applicant. When you have singled out one or two people, call each and every reference listed on the application. People sometimes aren't what they seem, and a few short phone calls may eliminate future headaches.

Be aware that what is said between you and the person giving the personal reference must be within the legal boundaries set by the federal government. This applies mainly to the person giving the reference, but every TIB will eventually find himself in this situation after hiring and firing only a few employees. In order to eliminate discrimination, remember that what may seem to be truth to you, may be slander in a

court of law. Consult your attorney or insurance agency for information on proper procedures.

Secondary Skills Are a Plus
When hiring new workers consider all phases of work done on the jobs, and not just the single skill that is immediately needed. Many employees have more than one skill or talent which can be developed and make that employee twice as valuable. Spend enough time with each applicant to determine all of his or her abilities.

LABOR LAWS

Every TIB who hires employees should be aware of the laws which govern the employer/employee relationship. The Department of Labor administers several laws which affect the operations of American businesses, both large and small. The sole proprietor, however, need not be concerned with any of them (except for minimum wage laws) until non-family members are hired as employees. At that point, the following laws apply to one's business.

The Fair Labor Standards Act of 1938, as amended. This law establishes minimum wages, overtime pay, record keeping, and child labor standards for employees individually engaged in or producing goods for interstate commerce, and for all employees employed in certain enterprises described in the Act, unless a specific exemption applies. Employers are required to meet the standards established under the Act, regardless of the number of their employees and whether they work full or part time. A complete copy of this Act, and additional information about the hiring of employees, can be obtained from your nearest Wage & Hour Division of the Department of Labor. (See "U.S. Government" in your telephone book.)

Occupational Safety and Health Act (OSHA) of 1970. This statute is concerned with safe and healthful conditions in the workplace, and it covers all employees engaged in business affecting interstate commerce and who have one or more employees. Employers must comply with standards and with applicable record keeping and reporting requirements specified in regulations issued by OSHA. Some states operating under OSHA-approved state plans conduct their own occupational safety and health programs. The law provides that a small business may request loans through the Small Business Administration when it can show that substantial economic injury is likely to result from a requirement to comply with standards issued by either federal programs or by approved state programs. In all states, businessmen and women may request the services of consultants who advise employers about compliance with OSHA, but who are not inspectors and do not issue citations for non-compliance. Priority for consultation is given to small business employers. OSHA training and educational materials and services may be obtained from ninety-one local agency area offices of the U.S. Department of Labor.

Social Security Act of 1935, as amended. This Act is concerned with employment insurance laws, and each state requires employers who come under its employment insurance law to pay taxes based on their payroll. For more information, contact your local Employment Security or Job Service Office, talk to an accountant, or refer to the Accounting section of this book.

Other laws administered by the U.S. Department of Labor. You need not be concerned with these laws unless you are involved in situations that concern 1) garnishment of employee's wages, 2) hiring of disadvantaged workers, 3) federal service contracts using laborers and mechanics, 4) federal contracts for work on public buildings or public works, 5) employee pension and welfare benefit plans, 6) government contracts, and other special situations.

FINDING EMPLOYEES' TALENTS

One of the most profitable traits a TIB will ever possess is the ability to assess a potential employee's character and talents before hiring, reassess those same qualities after hiring, making good use of all of his or her talents and abilities, and drawing out quality character traits. Placement and utilization of each employee is possibly as important as the actual talents of the individual.

Most workers are far more talented than even they re-

alize. Finding the right niche for each person can transform them from the average worker to the skilled, multi-talented worker. This will require proper training, but the time spent in training is what a TIB must give in order for any employee to expand, learn, and develop as a craftsman. Certainly you must give them the freedom to grow and express themselves through their work without becoming overbearing and bossy. But you must also spend the time necessary to work side by side with each employee until he has reached a certain level of expertise.

When evaluating employee abilities it may be necessary for you to expand your boundaries and ideas about how the business should operate and the kind of responsibilities the employee should have. In other words, there may be tasks that are outside the realm of physical labor that could be done by an employee with that specific talent.

For example, when gathering information about new materials or products and methods of use, look to the employees for a good source of information. Some individuals have a natural talent for finding out information on new products. When you find this talent, use it to its fullest. Get the employee involved in locating materials, gathering information, shopping for prices, talking with other contractors, etc. This natural talent can be a vital asset to the business.

HIRING FAMILY AND FRIENDS

No one can have too many friends, but close friendship with employees must be handled with care. When it is time to start hiring new employees, try to avoid personal friends or at least be advised of possible problems. With all employees there must be a business relationship first and friendship second.

In times of stress, the close personal ties with an employee may result in an emotional strain on the relationship rather than a typical business problem that could be left in the office at the end of the day.

Family and/or friends might assume that because of their close relationship with you they should have a voice in the business operations. They may feel the need for an explanation of your business decisions such as hiring and firing, duties of employees, handling money, raises, etc. The decision to hire a close friend, like any other decision, must be made wisely, taking into full consideration his character faults and strengths, personality traits, family relationships, and his capacity to benefit your business.

No one can give a set of rules on family and friend employees, so make these hiring decisions very cautiously. In some cases it works out, and in some it doesn't. Don't take a good relationship and destroy it by making a hiring decision in haste.

Working With Your Spouse

Who knows why hiring family members as employees seldom works, but the odds are against it. Possibly because family members usually do not have the normal consideration for each other's feelings and tend to speak sharp and sometimes cruel words to each other. We are sometimes more critical of family than we would be of any other professional relationship. This is especially and unfortunately true when a spouse works as secretary/bookkeeper. Husbands particularly do not treat their wives, who they say they love more than any other, as nicely as a casual acquaintance. Therefore, they are much more demanding and harsh when giving instructions to a spouse. A little patience and understanding can help a marriage that is already stressed by the heavy pressures of the business.

Hopefully, the relationship you have with your spouse is good enough that he or she can help with the paperwork of the business. In the early years, it is so much needed in order to keep the overhead down and to help speed up the paperwork process, so that there is time for family needs and recreation.

However, not every spouse has the skills to do the bookkeeping and secretarial work. Applying pressure on an unskilled spouse to fulfill these obligations may only bring guilt and frustration. Every individual is different. Their talents, desires, and abilities are completely different from everyone else's. Assuming that anyone can take on the office responsibilities is a big mistake. For those spouses who simply cannot or are not willing to provide these services, then the TIB must find alternatives. This is not a failure in a relationship, it is evidence that talents and skills lie in other areas.

EMPLOYEE INDEPENDENCE

A TIB who fears the success of his employees is destined for failure. There are some TIBs who are very concerned that their employees will become too independent or become better craftsmen than they are. As a result, through poor management, they prevent or indirectly discourage people from performing at their peak.

The goal of every business manager is to train employees to become winners. A manager must continually strive to help his employees develop skills that will improve their worth and bring independence. A good manager is concerned about the future welfare of his employees, about their ability to better themselves and work toward a more secure future.

At one time productivity was considered to be the single-most important gauge of management success. Many TIBs are realizing that dollars and cents are not the only way to evaluate success. More and more TIBs are defining success as how they build, train, and develop people. The future growth of the business is based on employee development. You want the employees to be achievers, and they will do so if they work in an environment which encourages self initiative and skill development.

People should be managed so they will begin to accept responsibility for themselves and their jobs. This means the TIB has to spend all the time necessary to help people develop the skills needed to consistently perform their jobs well. In addition, a TIB must build the employees' confidence and self esteem so that they can work independently of the TIB.

A good manager works to develop his people into star performers. He is not in competition with them. He's not afraid they will surpass him in ability. When they are performing at their peak, he should have a sense of accomplishment.

Personal development will eventually lead some employees to start their own businesses. This is a natural process, as they work toward making life better for themselves and their families. Help them along just as you needed help when you wanted to become independent.

EMPLOYEE DISMISSAL

Hiring is certainly much easier than firing. However, firing is a necessary function which the TIB must exercise from time to time. Most people hate terminating anyone for any reason. Employers who are not experienced managers tend to either dismiss someone in haste or go to the other extreme and let them get away with "murder" with little or no correction. Learning a balance will help the business run much more smoothly.

Use Your Head, Not Your Emotions
Obviously, if the workload has dropped off, the immediate thing to do is to lower the overhead. If there is not work to be done, then get the workers off the payroll. This can be the hardest decision you will ever have to make, but by unnecessarily carrying the overhead for an extended period of time, you could end up in serious financial trouble. As the need arises, begin laying off individuals who are not vital to the main functions of the business. If the workload situation continues, then you may have to lay off the balance of the workers.

Hopefully, their loyalty will cause them to stand by you until more work comes, but don't be afraid to lose them. It is better to train new employees than to lose all of your money by paying wages without any incoming contracts to cover the expense. Construction workers, especially, understand that this is the nature of the business.

Look Into the Situation Closely Before Acting
Lying, theft, drunkenness, fighting, taking kickbacks, or other immoral acts are certainly grounds for terminating an employee. Allowing this activity to continue without correction can spread ill will among the other employees. They might think that if "this guy can get away with and not get fired, then I can do it too." Of course, even with bad offenses, long-time,

faithful employees may need a little extra grace over the new ones. They have served you well and possibly deserve a second chance or at least a chance to provide an explanation for their actions. Sometimes personal or family problems can result in a change in an employee's behavior. His actions may be a result of frustration or desperation. A little understanding on your part may be just what he needs.

Many employers don't realize the great responsibility they have toward their employees. Employees aren't machines that can run without attention. They are individuals with families to support and bills to pay. If there is a serious problem in an employee's life, be understanding and compassionate, and when he gets through that problem, you will have gained a loyal employee.

Keep Within the Legal Boundaries
On the subject of employee dismissal, every TIB must evaluate his or her dismissal procedures. Educate yourself on employees' rights, and observe them strictly in order to stay out of legal trouble.

There is certainly nothing wrong with trying to treat employees as fairly as everyone else, and some TIBs may think that they are fair beyond reproach. However, they may not be acting entirely within the law. Caution must be exercised to prevent wrongful discharges. Laws can vary from state to state and from union to nonunion, according to the terms of agreement with an employee upon hiring, and according to the employee documents or lack of documents in your files relating to each employee. These documents will include needed information should you be legally challenged by an employee upon dismissal.

Your employees' files should contain such information as employment application, tax withholding forms, insurance information, absences, payroll records, written or oral warnings (and the presence of a witness at these times), or other records leading to dismissals. As always, don't just trust your own ethics, know your rights. These rights will vary depending on your location and particular situation.

If it is time to let an individual go, then make it a swift

and clean cut. There is no need to spend hours discussing the problem or justifying your decision. Don't let him stay on another day after you have told him that his services are no longer needed. He will be of little use, and may not be trustworthy after he knows he is let go. If you think he deserves special consideration, then give him a week's pay but tell him to leave immediately. It's better to send him on his way with pay than to keep him around for one more week.

Certainly all employees, even bad ones, should have the courtesy of being dismissed in private. There is no reason to publicly humiliate or offend anyone. Regardless of your dislike for this individual, he deserves to be treated with respect and decency.

FORMER EMPLOYEES

A good relationship with former employees can eventually bring good results. The former employee will take his respect and high thoughts of you to every corner of the business sector. Many will work themselves into prominent positions that can influence potential clients. Even employees who left on a bad note may have good things to say if you were acting justly with good and fair business decisions. Do not ever underestimate the need to continue a good relationship with every individual who crosses your path.

Just as you would have wanted your former employer to be understanding of your desire to start your own business, so must you have that kind of understanding. There will be times when employees will, if you have trained them well, want to venture out on their own. Help them out as you are able and avoid selfish attitudes toward this new competitor. By treating others fairly, they will return the favor. These kinds of relationships can develop into very productive and fruitful ones over the years. A former employee can be an excellent subcontractor or partner on large jobs when work is booming.

If you have to lay someone off during slow times, don't be embarrassed to ask them to come back when work has picked up again. If you have treated them fairly and they need employment, chances are they will not hold any grudges against you and will work

as well as, if not better than before.

ILLEGAL IMMIGRANT WORKERS

In many regions of the country the hiring of illegal immigrant workers has become a way of life. Small businessmen in particular have utilized these individuals in their businesses by employing them for every type of manual labor position. These workers has proved an asset to TIBs because of their low wages and hard work. However, they have become a serious problem to the American labor force by filling jobs that could be held by legal citizens.

In an effort to address this problem, the federal government passed a law known as the Immigration Reform and Control Act of 1986. This Act prohibits employing illegal immigrants. The purpose of this law is to remove the drawing power of available jobs from illegal immigrants by requiring employers to hire only U.S. citizens and penalizing those who hire illegal entrants.

Employment Eligibility Verification Form (Form I9)

Under this new law employers must require every employee to fill out the Employment Eligibility Verification Form (Form I9) within three days of the date of hire. This form requires the employee to provide a document or documents that establish both his identity and employment eligibility. The approved documents (U.S. Passport, Certificate of U.S. Citizenship, Certificate of Naturalization, state-issued Driver's License and Birth Certificate, etc.) are listed on the form. This form must be filled out be every employee except those hired prior to November 7, 1987.

Penalties for Not Complying

The Act provides for stiff penalties for nonconforming employers. Employers hiring or continuing to employ unauthorized employees may be fined as follows:

First Violation — Not less than $250.00 and not more than $2,000.00 for each unauthorized employee.

Second Violation — Not less than $2,000.00 and not

more than $5,000.00 for each unauthorized employee.

Subsequent Violations — Not less than $3,000.00 and not more than $10,000.00 for each unauthorized employee.

Employers who fail to complete, keep on file, and have available for inspection the Form I9 as required by law may face civil fines of not less than $100.00 and not more than $1,000.00 for each employee for whom the form was not completed, kept on file, or made available during a U.S. Immigration and Naturalization Service (INS) inspection.

Employers convicted of having engaged in a regular practice of knowingly hiring unauthorized aliens (after November 6, 1987) may face fines of up to $3,000.00 per employee and/or six months imprisonment.

Persons who use fraudulent identification or employment eligibility documents for the purpose of satisfying the employment eligibility requirements on Form I9 may be imprisoned for up to five years or fined or both.

As you can see, the federal government has implemented a law that has very serious ramifications for the TIB. What may have been an accepted practice of hiring illegal immigrant workers must cease. The INS is very serious about enforcing the Act and will steadily increase its enforcement procedures each year. The government has now taken the responsibility from the border patrol and placed it on the shoulders of the employer.

This has been a very brief explanation of the Immigration Reform and Control Act of 1986. To find out more about this Act and order detailed instructions for filling out Form I9 and additional copies of Form I9, write Superintendent of Documents, U.S. Government Printing Office, Washington, DC 20402, or contact the nearest office of the Immigration and Naturalization Service.

8. CLIENTS

SELLING YOURSELF

The quality of the work must be high, and the price must be right, but one element that a TIB many times neglects is the importance of "selling yourself." The Carnegie Institute of Technology analyzed the records of 10,000 people and arrived at the conclusion that 15% of success is due to technical training, brains, and skill on the job, and 85% is due to personality factors, to the ability to deal with people successfully. It is therefore understandable that the client places such importance on personality and character.

The client wants to know that you have the personality and character of an individual he can work with and trust completely. He may not know anything about you or your work, so in the initial meetings personality is extremely important. The first thing to remember is to be relaxed and be yourself. Trying to impress someone usually comes across as just that. There is no set formula for selling yourself. However, it will help if you keep these rules in mind when conversing with the client.

Avoid Negatives
Never put down the competition. Telling the client how bad the competition is will definitely be counterproductive, and will most likely drive him away. Sell the positive points of your work without giving much mention to your competitors' work. If the subject comes up, be diplomatic.

Make the Customer Feel In Charge
Let the client set the pace for conversation. Some TIBs, especially when they are tense or nervous, tend to be overbearing or talk too much. The client must feel comfortable with you, and this means that he must be able to think clearly and formulate his thoughts at his own pace.

Loud voices, finger-breaking handshakes, and hard back slaps are obviously out. No client, or anyone else, likes an overbearing, obnoxious person. He wants to feel confident that he can work with you during stressful times without serious conflict. Answer his questions accurately and concisely, always in a relaxed tone of voice.

Prepare for the Meetings
It is a must to be prepared for every meeting with a client. Preparation produces success. If there are prices to be obtained, details to be figured out, or paperwork to be done, always complete these tasks ahead of time. A prepared TIB will be thought of as a professional TIB. The client will be noticeably impressed by your skill and business control. On the other hand, the client will be very unimpressed by what may seem your lack of creativity and professionalism, if you arrive unprepared.

If the opportunity lends itself, try to schedule one of the initial meetings with the client at one of the job sites. Conversations about the workmanship become much more meaningful and understandable when you can look at the work. For this reason, keep the job sites clean at all times. If you know a client will be visiting the job, have it cleaned up before he gets there. However, trash and tools should be picked up and put in their proper places at all times, and not just when a client is scheduled to visit.

Make Your Meeting Top Priority
Be on time for every client meeting. Nothing aggra-

vates a client more than your being late for a meeting. Be early if possible. Use the spare time to inspect the work or to complete necessary paperwork.

Interruptions are a way of life for a TIB, but don't allow a meeting with a client to be interrupted. Most problem situations can wait until the meeting is over. If an emergency arises, excuse yourself politely and suggest that the client walk through the job or review material brochures. Take care of the problem as quickly as possible and return to the client with your apologies. For this reason, an employee trained to handle certain problems in your absence is a necessity.

Don't Feel Pressured

The nature of a client is to want pricing information as soon as possible if not immediately. Feeling the pressure of the situation, a TIB may quote rough prices without giving it proper consideration. This can be very risky. Quoting prices should not be done casually. The client will expect you to be an authority on the subject. If the quote is completely off base, the client may begin to question your knowledge and reliability. In addition, the client will question your ability when an "off the top of your head" quote is given, but then you quickly change your mind showing confusion and uncertainty. If you are unsure of the prices, the best thing to do is to tell him that you will have to get back to him later with the price. Don't feel compelled to have this information at your fingertips. However, if your work is done on a per unit basis, such as $2.50 per square foot, you should be prepared to give pricing information to the client relatively quickly unless the job is unusually complicated.

Be Interested in the Client's Personal Interests

The purpose of any meeting with the client is to discuss some aspect of your work and the possibility of a future business relationship. However, clients are individuals with likes and dislikes, hobbies, goals, families, abilities, and interests. They, many times, enjoy discussing something other than business. The ability to talk to others on different subjects will help transform a business deal into a business relationship.

Find out the client's interests, stay in tune with current events, be interested in him as a person and not just as another job opportunity.

It is amazing how many contracts are awarded as a result of a meeting where everything was discussed but business. If the client likes you personally, he may hire you regardless of other considerations. That's not necessarily the way it should be done, but it shows the importance of good conversation and personal character.

Satisfaction Guaranteed

Something should be said about customer satisfaction. If you live up to the old saying, "Service after the sale," you will have a good chance for customer satisfaction. A satisfied client is the best advertising possible. Knowledge of your work will travel by word of mouth from client to client, which makes your job of selling much simpler. The client could be sold and ready to hire you even before you meet. It's worth repeating: there is no better advertising than a client who is completely satisfied...and there is no worse advertising than a client who feels he is not getting the proper service.

A good way to assure the client of your interest in his future business and gratitude for the work is to send him a letter of appreciation (Figure 18) after the job has been completed. This is seldom done in the construction business, but the results will be remarkable.

Business Productivity

While the TIB's business must look orderly, it must also act in an orderly manner. It must do things in a predictable, uniform way. Regardless of what service is available to the client, your business must perform that service with the same speed, quality, and service the tenth time as you did the first. If the expectations created in the first job or two were violated during each subsequent job, the client won't know what to expect. The unpredictability says to the client that the TIB is in control and not the client. He must feel that you are running your business for him and not for you. The services provided by the TIB are not nearly as important as the consistency with which they are performed.

Clean Up Your Act

Selling yourself to the client will be hard enough without your doing something that will automatically irritate him. Without exception, every client has certain pet peeves, things that happen on jobsites from time to time that create a progressive level of irritation. You won't always know what will set off a client, but here a few things that should definitely be avoided.

Children

Don't bring your kids on the job. Sometimes you may not have a choice because of a personal situation that requires the children to be with you all day. Just be aware that they must be watched and under control at all times. The client will not appreciate walking through the job and suddenly finding a couple of kids playing in one of the rooms. It isn't professional, and he will resent it. While he is inspecting the work or meeting with the other workers on the job, he doesn't want to have to worry about what someone's children are destroying. And he certainly doesn't need the risk of your kids being injured on his job. Children do not know how to watch out for dangerous situations, and few places are more dangerous than a construction site.

Pets

Dogs can also be a major source of irritation for a client and should be left at home. Unless they are chained up away from the job and its activity, they can cause costly damage. You've lost all hope of getting future work from a client after your dog just messed on his newly laid hardwood flooring. If the dog is mean, don't even think about taking him to the jobsite. The chance of his biting someone, possibly the client, is far too great and not worth taking. The client will not appreciate this unwanted element in his daily routine and will have less regard for you as a result.

Friends

From time to time it's nice to be able to show off your work to a few friends. Unfortunately, the client may feel somewhat protective about his job and might not appreciate a group of tourists marching through his project. If you want to visit the jobsite with a group of people, ask the client's permission in advance. As much as you would like to show off your work, visit-ing the job should be held to a minimum.

Drinking

Probably the most common irritant for any client is when a TIB or his employees drink alcohol on the jobsite. What your employees do on their own time is their business, but as long as they are on the jobsite, you still have authority over their actions. Even through the client may have a drink on the jobsite, he may not think very much of your drinking on his job. He's the boss and can do anything he wants to, including having a drink. This does not give you the freedom to do the same. Without exception, you should never join the client in a drink on the job. He will respect your policy and admire your integrity.

Parking

There is another thing that some thoughtless TIBs do that creates an inconvenience for everyone. Most jobsites have several workers with several vehicles. In addition, there is usually a shortage of parking spaces. When you or one of your employees show up on the job, find a parking space that is away from the main traffic flow. Delivery trucks, other contractors, and the client must all pull up to an unloading area. If your vehicle is parked in the way it will be aggravating for him to come find you and wait until you move. Find a space far away from the main activity and leave the good parking areas for the client and his employees.

CLIENT ENTERTAINMENT

Client entertainment is often misused. Some TIBs try to do what is unnatural for them by trying to impress a client with expensive and elaborate entertainment. There is no need to attempt to buy the work. Whatever is done must be natural and within the TIB's financial limits.

The entertainment you involve yourself in must always be consistent with your lifestyle and values. Conservatism, family orientation, and respect for others are true signs of strength of character and stability.

Lowering your standards to join a client in activities you feel are wrong will put you in a lesser light in the client's eyes, even though he was the instigator of the activity. Most individuals know the difference between right and wrong. The client will have less respect for you even when he himself was involved.

There are times, however, when discussing business over dinner, going to a sports event, or inviting him on a hunting trip can help promote a good business relationship. Be wise in both word and deed, and you can avoid some unpleasant and regrettable situations.

DRESS AND APPEARANCE

Appearance is very important. If you overdress on the job, you can be quickly pegged as one who has lost touch with his craft. Remember, you are a working TIB. Even partially overdressing will keep you from demonstrating your expertise when on the job. The key is to dress neatly and cleanly, and don't worry about a little honest sweat. When visiting a client's or banker's office, it is especially important to dress neatly, but it is acceptable to dress in your everyday work clothes. A hard worker soiled in what is easy to see were clean, neat clothes that morning will win out over a dude in a suit. However, mud on the feet, soiled chairs, and dirty hands on furniture will cause you to be quickly disregarded, so use a little common sense.

Keep breath mints handy for those unexpected meetings, but don't roll the mint around in your mouth while you are talking. Keep your hair neatly trimmed and wash it regularly. Fresh dirt is not a problem, but bad personal hygiene is a real turnoff. If you didn't learn good hygiene in early years, now is the time to make the change. Observe others for tips and work diligently at improving your appearance.

COMMITMENT TO OBLIGATIONS

"If you can't do it, don't say that you will, and if you said it, then do it." That statement may seem overly simple, but it's the exact truth. Lack of commitment to obligations is a great tragedy in today's workforce. There can be all sorts of reasons to back out of a commitment. Losing money on the job has probably been the biggest semi-legitimate reason for backing out of a commitment. However, you must finish what you

start and finish it with quality workmanship and in a timely manner no matter what the money situation. Obviously, if the loss is great enough to threaten the stability of your business, drastic measures may be needed, but hopefully this will not be the case.

Many deals have the potential to be broken or voided because they were not legally binding for one reason or another. The fact that legally you don't have to pay someone or complete the project or fulfill an obligation does not release you from a moral and ethical obligation. You must stand by your word as a professional. If you said that you would do a job in a certain way, then you are obligated to do it just that way, whether it is in writing or not.

Too often TIBs and clients alike have taken the easy way out of a situation by doing only what was legally required. Do not ask yourself, "What do I legally have to do?", but ask yourself "What did I say, or imply, that I would do?" Initially, it may seem harder to do what is right, but it will bring long term rewards. The hardened conscience of a man who has continually escaped obligations is something you never want to have.

PROBLEM SITUATIONS

The ability to relate easily to the client so that you get what you both want and make him look good is worth its weight in gold. Some clients are experienced individuals with great personalities and are enjoyable people to work with. However, there will be those who, for one reason or another, are very difficult to work with. Learning to handle problem people and problem situations will project a professional image.

Most people don't want to be difficult, they just want to get a certain result. Unfortunately, many individuals have learned only one way to get what they want — by force. Your reaction will either remedy the situation or make it worse. In some cases with some people, a firm word and a show of strength will work wonders, but in many if not most situations, a patient and calm response from you will produce the same effect. Reading the client can be a tough task which becomes easier with experience.

Honesty, The Best Policy

When working through a problem situation with a client, don't make the mistake of thinking you have to have all the answers. Certainly, you must know your business and be prepared and knowledgeable about the job and its details, but if you don't know something, or if you have done something wrong, it is best to admit it. Honesty in a situation will mean more to a client than proving you were right, especially if you lie or tell a half-truth and he finds out later that you were not being straight with him. That will be devastating.

It's easy to tell a half-truth. For example, there he is, ready to pounce on you for making a mistake. You realize the blame can easily be put on someone else which will help you avoid the next traumatic thirty minutes. It's so easy — but don't do it. Stand up for what is right and fair, and you will find that the client will not be nearly as angry if he knows you are owning up to your shortcomings. People want to trust and be trusted. To admit, "I was wrong, I am sorry. How can we work out a solution?" usually takes the punch out of a touchy encounter.

There is seldom a situation so bad that, with proper negotiating, it cannot be resolved. Unfortunately, many problem situations on the job could have had a relatively easy solution, had the emotions of both sides not been so out of whack. The personalities get tangled in emotion, and what was once a simple problem reaches an impasse.

Handle each problem calmly and creatively. If it is your fault, fix it. If it is not your fault, don't be pressured into spending money to save the client a buck. If it's relatively inexpensive to fix, it might do you well to fix it whether it is your fault or not. This will go a long way in client relations. Be a professional in the discussion, and remember that there is a workable solution to every problem.

Eliminating Bad Clients

A tradesman starting a new business needs all the work he can get his hands on. However, from time to time, as anyone with a few years' experience will tell you, there will come along a client that you should never get involved with, or a current client gone bad who must be eliminated. Unfortunately, not everyone is honest and fair, and as a result the TIB must always be on the alert. The ways for an individual to be a bad client are numerous. Basically, the problem is that an individual's character is such that he cannot be trusted, and you fear not getting paid in full after the work has been completed. Reading a person's character is very important. You can't believe everything a potential client tells you.

Avoid persons who are known to be dishonest or that you suspect are dishonest. There are people who will take you for all you're worth and will have no regrets or remorse. There are plenty of honest people in this world with whom to do business, without having to put up with liars and cheats.

Beware of Con Artists

Watch out for individuals who talk a good story but are basically unknowledgeable and unsuccessful. They may or may not have good intentions, but if they don't know what they are doing they can get themselves, and you if you let them, in a serious money bind. Around every corner is a guy who is trying to pull a deal together. Usually, he doesn't have the financing and is looking for a pot of gold at the end of the rainbow. He is usually unskilled in how to get his "deal" off the ground, and will use you to educate himself and open doors with others. He might ask questions about how others get financing for projects or where you find most of your work. This type of person is looking for a get-rich-quick deal, instead of working hard and building slowly. This person will waste hours of your valuable time, if you let him, without any return to you.

On the other hand, intelligence and know-how are not signs of character either. The most intelligent person in the world may be a rat or a jerk if he doesn't have the inner convictions of honesty and integrity. Don't be taken by a fast talker who shows a good measure of knowledge, but has no proven track record. Try to check out his references whenever possible. Ask other tradesmen about their dealings with him. Check with suppliers for possible references. Be cautious and use good common sense.

Know When to Call it Quits

Now comes the tough part. Let's say you have a client who has paid on time and fairly for several jobs. Suddenly, he is slow in paying on a job or two. The payments become further and further past due. What do you do? Cancel him as a client and lose his business, or continue to work for him until he gets his feet back on the ground?

Both decisions have equal merit. If you suspect serious financial problems, sometimes the best thing to do is to cut your losses and avoid losing any more than necessary in case his business fails. On the other hand, if you hang in there with a good and fair client, he may be able to pull out of the slump, and your relationship will be even stronger for having stuck by him through his tough times. The decision is up to you. Weigh all the facts carefully. Try to get as much true and factual information as possible about his financial condition before making the decision.

Many times, the choice is made automatically by your inability to carry the job financially until he can pay you. Remember this, not doing anything and ignoring the situation is a decision, a wrong one, but nevertheless a decision. Make your decisions timely, with discipline, intelligence, and direction.

An alternative to cutting off a reliable client in financial distress might be to put him on a cash basis. This can be accomplished by requiring advance payments before the work with progress payments each week. The contract covering the work should be carefully drawn to show this type of agreement. In addition, if materials must be ordered for this work, the client should pay for these in advance of ordering them. Because of his financial condition, you may have to order materials on a weekly basis to keep from tying up his operating capital. This procedure entails extra effort on your part, but it is an option to consider to help out a client in financial trouble.

RELATING TO THE SUPER RICH

The way to act with the super rich, super intelligent,

super powerful, or super anything is to treat them like anyone else. Of course, you must recognize that their position may not afford them much leisure time, so don't waste it with idle talk. These people generally want to act like normal people and to be treated as such. They don't want you to make an issue of their money or position or try to impress them with yours. Just relax and be yourself.

As a word of caution, wealthy people generally have lots of possessions and properties. If you ask to fish in their rivers, swim in their pools, and hunt on their land, you may become a real nuisance. They need privacy like anyone else. They may enjoy inviting people over or letting them use their property, but be careful not to abuse this privilege. Certainly if this individual is your client, overstaying your welcome jeopardizes your business relationship.

Occasionally there comes along a person who wants the world to know how great he is and will enjoy walking all over people less fortunate than he. Unfortunately these rude, but wealthy people lack the character and integrity to treat everyone fairly no matter what their financial position. We are all talented in one way or another, and just because one person has more wealth than another does not mean that he is more or less talented. He is simply more fortunate and should treat his position with gratitude and humility, not pride and arrogance.

Many times the super rich have sacrificed and worked extremely hard for years to achieve success only to find that their sacrifices included family, friends, and happiness. They may not be any more fulfilled than the next guy, because their personal priorities were out of order. These people are not idols and may or may not be worthy of our admiration. Look at what kind of man or woman they are and base your respect on their character, not their possessions.

CLIENTS' EMPLOYEES

A terrific way to please the man in charge is to please his employees. The client's best and most valued source of information is these individuals. In fact, they may be the ones who have more to say about hiring a TIB than the boss. Therefore, developing a good relationship with the client's employees can be crucial.

Getting acquainted is certainly in order, but be cautious not to let your discussions take too much time for their duties during work hours. You don't want to further a problem that the client may be trying to correct such as laziness or lack of discipline. Furthermore, you certainly wouldn't want to keep an employee from completing a task that needed to be done by a certain time.

Eventually, you will run into a discontented employee of a client — one who is unhappy with his job and openly puts down his boss. Be careful not to involve yourself in a conversation dealing with the client's faults. One way or another, maybe because that employee is a two-faced creep, this information can get back to the client. In other words, even if the client is a jerk, don't discuss it with his employees. Keep a neutral position and positive and enthusiastic frame of mind at all times.

You should seldom inform a client of a bad employee unless the offense is extremely serious such as violence, theft, etc. Avoid informing the client of a bad employee attitude, or that he left work early. During the disciplinary process, the client may tell his employee where he got the information that led to the disciplinary measure. As you can see, the employee will be very difficult for you to work with from then on.

Help, But Don't Hire

Hiring the ex-employee of a client can be a big mistake even if the client gives his blessing. In many cases something went wrong and the relationship went sour, resulting in the employee's departure. Even if the employee was completely in the right, it could constantly remind his ex-boss of past problems or his own shortcomings. In addition, the ex-employee may have privileged information about the client which might make the client uneasy about con-

tinuing his business relationship with you. This is not a hard and fast rule, and obviously many employees leave their employers without any continuing problems. Still, it would take a unique individual to be able to make a transfer without some measure of adjustment.

As these employees leave the client, however, you may find it beneficial to help them along. This can serve your business very well. It is a good idea to send a letter of congratulations (Figure 19) soon after he takes the new job. It's likely he will find employment with a similar firm, and if your relationship is good they may bring you into a working relationship with a new client, maybe even doubling your workload. Never burn bridges or step on people. Help those you can with a little kindness and support.

Be Honest and Aboveboard

Never, ever give or take kickbacks, payments, or gifts for work. It is unfortunate but some people, including employees of the client, will try to make a few bucks by offering you a deal. In other words, if they get the client (their boss) to accept your bid (padded with kickback money), then you are to give them a few unmentioned bucks. Run, do not walk, away from such a situation. Try to politely but firmly turn this offer down. If you value the client's work, there may be an opportunity to inform him of this activity. Don't blow the situation out of proportion, but explain exactly what was said.

This sort of thing is as dishonest as you can get. If you ever get involved in this kind of "special deal," it will come back to haunt you in the future. The best policy in any business deal is for everyone involved to know the basic agreement and the responsibility of each individual. There should be no hidden agreements or payments. Honesty and integrity may not always be popular, but handling your business in this way will help you sleep at night, and in the long run will prove to be worth their weight in gold.

9. SUBCONTRACTORS AND SUPPLIERS

COMMITMENT TO OBLIGATIONS

Most TIBs cannot successfully operate their businesses without continuing support from material suppliers and, in many cases, subcontract laborers. Developing good long term relationships will prove beneficial many times over.

Material Suppliers

With material suppliers the TIB's main objective is to be a reliable and trustworthy customer. This is accomplished in several ways, but the single most important way is the most simple — pay the bills on time. There is nothing in the eyes of the suppliers that turns a TIB into a second rate customer faster than having overdue invoices. It is wrong and unprofessional, so don't do it if you can in any way avoid it. Paying the bills late tells them you don't care, and it shows how willing you are to take advantage of them. Basically you are asking them to fund the project. These companies have bills to pay just as you do. When you are regularly late on payments, what do you think the supplier tells his other customers about your business? Word spreads fast...incredibly fast.

Something to remember about paying invoices for material purchased is that most suppliers will provide discount prices if the invoice is paid promptly. This discount usually ranges from 1 to 2% of the total bill. This may not seem like much but can really add up over time. It's money in the bank, and its loss is a terrible waste simply because the check wasn't sent a few days earlier.

Subcontract Labor

Like material suppliers the obligation to the subcontract laborers is a serious commitment; one that cannot be taken lightly by either the TIB or the subcontractor.

The sub must perform as agreed and by doing so expects the TIB to do the same. The TIB's failure to pay the sub's invoice by the scheduled date is just as serious as if the sub failed in his responsibility to perform the work as originally agreed. There is no difference.

The subcontractor is himself a TIB. He expects to be treated just as you do. Many TIBs cannot afford to carry unpaid invoices for any length of time. Not sending him the due payment may be more than an inconvenience. It may put him in serious financial trouble. Every TIB has the moral obligation to treat others fairly and honestly. Paying the bills late shows neither.

What to Do if Things Go Bad

Even with the most dedicated TIB a bad economy or other unforeseen circumstance may cause a bill to be paid late. During this rare time in your business there are specific steps that must be taken. The first thing to do is to contact each and every creditor and patiently explain the situation. Don't wait for them to call you. It's your job to contact them. Face the problem head on. They will probably understand the problem, but they will want to know the exact date they will be receiving the check. Tell them, if possible, but don't lie or give false promises based on hope and speculation.

If you don't know when you will be able to pay, tell them you don't know. Try to persuade them to stick with you during this hard time and assure them of your commitment to fulfill your obligations. Contacting them first will help convince them of your honesty and concern for the problem.

If for some reason payment is based on certain variables, keep them informed on a regular basis as to the status of the situation. It's their money, and you owe them your undivided attention and assistance.

INFORMATIVE RESOURCE

A TIB must keep himself informed of new products, applications, jobs out for bidding, available clients, etc. Excellent aids in gathering this information are subcontractors and material suppliers, via a good working relationship.

Never underestimate the value of the subs' information. These specialists can keep you abreast of new job developments in the area. They know much more than you may realize. However, you should be careful to respect their need to protect their other clients just as you would want them to protect you. Grilling a sub for information may make him feel uneasy and pressured.

Most material suppliers are in constant contact with manufacturer's representatives, new products, and product literature that keep them in touch with the everchanging construction industry. Because of the constant change and advancement in technology, not every supplier is able to keep up with all of the available products. In fact, some are and will continue to be useless in regard to what is going on in the world around them. Therefore, make contacts and establish credit accounts with a variety of suppliers. If you have to purchase a specialized product for a particular job and one supplier does not have it or can't get it, then it's a good idea to have credit with a variety of suppliers.

The difference between the TIB who was awarded the contract and the one who wasn't is sometimes the knowledge of materials, practices, and application procedures of new products and technology. New methods are out on the market every day. Some work

and some don't. It's the TIB's job to find out as much as possible about each one, and the suppliers are the closest contacts with these new products.

HIRING SUBCONTRACTORS

The TIB will not require the services of other TIBs/subcontractors until the business expands to the point that the scope of work involves other trades. Many TIBs realize that these specialists can do many of these jobs faster and better or at a lower cost than their own men. The decision to have a subcontractor for a particular job is made after careful evaluation.

Subcontractors should be hired as carefully as the general manager of a professional football team selects his players. Each sub must be carefully evaluated based on his total effect on the TIB's performance for the client. Successful evaluation of a subcontractor is first done through careful explanation of the sub's responsibilities. An open and detailed dialogue with each sub will reveal his knowledge of the specified work. This discussion will also give the TIB insight into his character and ability to work with people.

When hiring a subcontractor there are specific areas of concern and questions that must be answered before an agreement is made.

1. How long has he been in business? This won't tell you much about a sub except that if he has been in business for twenty years it can be assumed that he knows what he is doing. However, if he has been in business for less than a year he may not have the experience or know-how needed to do quality work. The length of time he has been in business is not a true measure of his ability, but it certainly lets you know to be very careful before hiring a possibly inexperienced sub to do the work.

2. Can he properly manage the job? The subcontractor must have a sufficient understanding of the scope of work to adequately supervise his portion of the job. If he doesn't seem to understand the work in the preliminary discussion, don't assume he will catch on later.

When a subcontractor's workload is extremely heavy, he may not be as available as necessary to watch his workers if he is busy taking care of other jobs. It is a good idea to find out about his current workload so that you can be confident of getting his undivided attention.

3. Is he reliable and prompt? If the sub does not show up on the job when scheduled, the TIB's ability to complete the job on time can be severely hindered. The TIB can find out about the sub's reliability from his past customers. They should be able to give details of his past performance. This is a question that should also be asked directly to the subcontractor. Evaluate his response and write it down so that you can hold him to his commitment.

4. Has he done this type of work before? The sub may have an excellent track record, but his workload may have consisted of smaller, less complicated jobs. For instance, a sub may do perfectly adequate work on a small residence, but he may have no experience on large commercial jobs. As you check his references, be sure to ask his customers about the type of work and the size of the job.

5. What size crew will be working on the job? A frustrating problem when hiring subcontractors is having an inadequately sized crew show up on the job. Promptness is critical with both starting and completing the job. If the crew is too small, the job will not be completed on schedule. The sub should make a commitment that a certain number of workers will be on the job at all times. Of course, the crew size requirement may vary depending on the stage of the job and emergencies that every sub has from time to time. Have a clear understanding of his intentions.

6. What kind of warranty does he provide? If a TIB warranties his work to the client, he should expect the same kind of warranty from the subs. Unfortunately, warranties are seldom better than the subcontractor's reputation. Warranties can and should be identified in the contract or proposal, but the best way to find out how a sub performs on warranty work is to contact his customers.

7. Does he have the necessary tools and equipment? Most subcontractors have the tools to complete their required tasks. However, if special or expensive equipment is required it is important to find out what he is planning to use and where he plans to obtain the equipment.

8. Does he have quick access to materials? A vital key in the subcontractor's ability to perform is his quick access to the required materials. Some subs will provide labor only which relieves them of this responsibility. However, for those subs who provide labor and materials, the TIB must find out the name of the supplier so that a credit check can be made. This becomes a problem for the TIB when a sub has trouble paying his suppliers. They may put the sub on a "cash only" basis, and if he doesn't have the cash to purchase the materials, he cannot work on the job. His optimistic hopes of scraping together the needed funds may fall through the day he is to show up on your job. A phone call to his suppliers will let you know his credit status. If he is in trouble, other arrangements should be made either by hiring another sub or purchasing the materials separately and hiring him for the labor only. Be careful when doing this if you are not very familiar with that trade. You can end up paying much more for the job if the material costs are much higher than the sub originally anticipated.

SECTION IV
SHARPENING THE EDGE

10. ACCOUNTING

EXPLANATION OF RECORD KEEPING

Keeping the books is one of the most misunderstood, confusing, and hated tasks for many TIBs. The word "bookkeeping" has turned the mightiest men into marshmallows. Because they don't understand it, they often end up ignoring it completely. Understanding and keeping accurate "books" is difficult because everyone, including many accountants and the federal government, have made it hard to understand.

This section will not deal directly with accounting or accountant-style bookkeeping. The explanations and examples used in this section will deal more with record keeping than with bookkeeping. Record keeping is a more accurate term to describe what the average TIB must be responsible for. Record keeping is the daily function of organizing and recording various transactions. Bookkeeping is taking the records that the TIB has kept and applying a few accounting procedures to them. An accountant can provide this function as the need arises. This will be once a month for some TIBs and less often for others, as per the accountant's recommendations.

It's Easier Than You Might Think
To start with, you don't have to be an accountant or have an accounting degree to keep the records. You can learn with a little common sense enough of the step-by-step procedures to master the day-to-day details without a single college class. Furthering your education is very important and should be a part of your long term goals, and classes in accounting and business would be very helpful. This is just to say that it is not as hard as you might think.

Like anything else, record keeping is a learned process. Just like learning a specific trade, you start by understanding a few basic principles, and the rest is primarily mechanical.

In today's economy, there are thousands of self-employed tradesmen. Most of them are technicians. They know their trade, but many of them don't know the various aspects of running and managing a business. The key is turning that technician into a manager. An important step in that process is the working knowledge of the record keeping system. Even though you might not be the person who actually organizes the bills and keeps records, it is necessary for every TIB to understand the procedures. A TIB must stay in touch with every aspect of the business. It will also help you understand what that fast-talking CPA has been saying.

What is Record Keeping?
Record keeping is a systematic recording of business transactions. Paying bills, receiving money, and paying employee wages are a few of the different types of transactions that regularly occur. When each check is written, it is applied or allocated to one of several categories. In other words, after a month's or year's worth of bill paying, you could place every single expense in one of fifteen to twenty categories that apply to your business. Keeping records is simply applying every check written to one of these different categories. At the end of each month, a total is determined of how much was spent for each category. It's like a detailed extension of your checkbook register.

So, why go to this trouble? First, at the end of each year your accountant can take the totals from each category and figure the taxes. Without this information you could pay thousands of dollars in taxes not actually required, or you might end up paying less than is required. Rather than face stiff penalties, the better alternative is to learn a record keeping system.

Regarding taxes, there will be several payments due to the federal government for payroll taxes throughout the year, and your record keeping system will keep track of what and when they are owed.

An accurate record keeping system will help the accountant determine whether you are making money or losing money. If you rely on the checkbook balance to determine whether you are making a profit, it may eventually put you in serious financial trouble. The checkbook balance will never, ever give you a true picture of how much money you have made.

From time to time you will be required to provide a (reliable) financial statement to your banker, bonding company, private investors, etc. Without an accurate and complete record keeping system it will be relatively impossible to produce this information.

Perhaps the most important benefit is the ability to capture accurately the actual costs on a particular job. This verifies whether or not a profit was made on that job. It also simplifies bidding similar jobs. Without a job cost record it will be difficult to determine which jobs made a profit and which were bid too low. This record can help you evaluate and adjust future bids.

In addition, the overhead expenses can be accurately reported, giving a true picture of what it takes to operate the business. So many TIBs have very little idea of what it costs them to operate their businesses, and most are shocked when they find out the actual yearly overhead expenses.

This is not an accounting manual and will not cover all the areas of bookkeeping that a TIB must eventually learn. The procedures given in the book are for the purpose of involving the TIB in the initial stages of the record keeping system. This information will help him establish a system for the day-to-day operations. Once these procedures have been learned, he will be able to educate himself, with some help from others, in other areas of bookkeeping. Eventually he will gain a broad understanding and proficiency.

Reference Material

There are a variety of useful books and guides that explain accounting procedures in a simple and easy to understand format. Refer to the Resource Materials section for a list of some of these publications.

TERMS AND DEFINITIONS

Before getting to the step-by-step procedures, a few terms that are commonly used in accounting and record keeping will be helpful. Only a few of these terms will be used on a regular basis by the TIB. However, a basic understanding of them will aid in successfully communicating with the accountant.

Accounts Payable — Amounts owed to others. The sum total of all invoices or bills from subcontractors and material suppliers, etc., that have not yet been paid. When each invoice is paid, it is then considered to be "Accounts Paid."

Accounts Paid — These are the paid invoices of what used to be "Accounts Payable."

Accounts Receivable — Basically the flip side of "Accounts Payable." This is money that is owed to you. Primarily it consists of payments due from clients for your services or materials, services that have been billed but that the client has not yet paid for.

Amortization — The systematic reduction of an account balance, such as a mortgage, over a specific period of time. Amortization is the tool used to determine the gradual reduction of the principal while making loan payments.

Appreciation — An increase in value. If a piece of real estate increases in its worth, it is said to have appreciated in value.

Asset — An asset is something owned by the TIB that has a positive cash value. Assets can be such things as real estate, equipment, accounts receivable, cash, etc. Assets offset liabilities on the Financial Statement (Figure 14).

Audit — The detailed examination of all accounting records to determine their accuracy and conformance with the law and accepted accounting practices. A CPA can perform an audit for the TIB if either the TIB or another party needs to know if his financial information is correct. Of course, the IRS uses the audit as their means of determining whether the correct taxes were paid.

Balance Sheet — A report of a company's financial position. It is sometimes referred to as the Financial Statement (Figure 14). It is a recap of all the financial information showing the assets, liabilities, and owner's equity.

Bankruptcy — A situation where the company's liabilities (what it owes) far exceed its assets (what it has to pay what is owed). The condition under the Federal Bankruptcy Act in which an individual's assets are taken over by a federal court and sold to pay the creditors. Bankruptcy may be voluntary (decided by the TIB) or involuntary (forced by creditors).

Budget Variance — In job cost accounting it is the difference between the bid (budget) and the actual cost of a job or a particular phase of the job.

Capital — When you first go into business you will make an investment which may consist of cash and other assets. This investment is called your capital. You may increase the capital by making a profit or by adding funds to the business. You may decrease the capital by withdrawing part of your investment or by suffering a loss. Basically it's the amount of money you have to operate with.

Capital Asset — A long term asset such as equipment or building. Usually a capital asset is of larger value than hand tools or small equipment.

Cash Disbursement Journal (Figure 20) — A special journal used for recording outgoing cash transactions

of the business. This is similar to a checkbook register. The main difference is that the cash disbursement journal allows the TIB to group every expense into a specific category. This type of journal can be used to write the check and categorize it at the same time.

Cash Flow — While the TIB's business may have a positive net worth, it may have a cash flow problem. Basically it becomes a problem when the bills for a project must be paid prior to the TIB being paid for the work. If there is not enough working capital (cash) in the account to cover the bills until the payments are received from the clients, there will be a cash flow problem.

Cash Receipts Journal (Figure 31) — A special journal used for recording all cash received. The cash receipts journal should list the name of the source of the cash, the amount received, and the purpose for which it was received (such as payments on services rendered, material sold, refunds, investment capital, etc.).

Chart of Accounts — The various categories that are used in the cash disbursements journal are known as the chart of accounts. All of the business expenses are identified by one of these account names.

Check Register — This is the record of all checks written by the TIB. The cash disbursements journal (Figure 20) is an excellent method of writing and recording the checks.

Credit — This can be very confusing. In accounting the term credit does not mean the same as it does used in everyday conversation. Credits have to do with making entries in the general ledger. Credits add value to some accounts and decrease others. This depends on the kind of entry and the type of account. Your accountant can help make the credit entries.

Debit — Similar to a credit, debit has an accounting meaning different than you might expect. It has to do with entries into the general ledger. It increases the value of some accounts and decreases the value of others. Your accountant can help make the debit entries.

CASH RECEIPTS JOURNAL

Page __1__ of __1__

DATE REC'D.	DESCRIPTION	AMOUNT
xxxx	GBJ Builders, 129 Peach, Invoice #919	4500.00
xxxx	Refund on utilities	100.00
xxxx	Bolder Const; Residence, final draw	2500.00
xxxx	Sold compressor and hoses	475.00
xxxx	B. Jones, Jones building, Invoice #1019	750.00
xxxx	J & K Construction, BFK Building (partial payment; balance in five days)	1000.00
xxxx	Wall Enterprises, advance pmt. on City building job	500.00
xxxx	Wall Enterprises, Invoice #1021 (less $500.00 advance)	1896.50
xxxx	Refund on returned materials, Jones building job	96.50
	TOTALS	

FIGURE 31

Depreciation — This is a form of spreading out the original cost of an item (such as a vehicle) over a period of years. In other words, in order to get the best possible tax deduction on the purchase of equipment, the accountant might set up a depreciation account in the general ledger so that the tax deductible expense effectively offsets your taxable income for, for example, five to ten years.

Direct Cost — Some costs of running the business are not job-related costs, but the costs that are directly attributable to completing the project (such as materials and subcontract labor) are known as direct costs.

Discount — Some material suppliers will give cash discounts if the TIB pays the bill by a certain date. This amount is usually 1 to 2% of the total invoice amount.

Equity — This is also known as net worth. Basically, as the TIB is the owner of the business, the net worth of the business is his equity. If all of the bills were paid and the equipment and property sold, the remaining cash value is the TIB's equity.

Financial Statement (Figure 14) — A report of the financial state of the business. It lists the assets, liabilities, and owner's equity. It establishes the business's net worth. It is formulated using the Balance Sheet and Income Statement.

General Ledger — As the payroll journal, cash disbursements journal, and cash receipts journal are filled out, the many entries in these journals will be totaled at the end of each month. The monthly totals are entered into the general ledger. Initially, this process can be done by a bookkeeper or accountant. However, as this monthly technique is learned the TIB may want to save money and handle the process on his own. Very simply, there are two types of entries: monthly (payroll deductions, cash received, and cash disbursed) and periodic (various items such as inventory, depreciation, insurance, etc.). The general ledger is very important, as it is used at the end of the year by the accountant to complete the tax return and other financial reports. This book will not go into a detailed explanation of the general ledger, so con-

sult your accountant for more information.

Gross Profit — Take the total amount received for services or materials sold, subtract the actual cost of labor and materials used to complete the work; the balance is gross profit. In other words, the profit before allowing for overhead expenses and other non-job-related expenses such as secretary's salary, office expenses, accounting fees, advertising, etc. See *Net Profit*.

Gross Sales — The total amount of sales. All revenue from work done or material sold before expenses and deductions are considered.

Income — Total money received that increased the business value. Similar to net profit but it includes all money received and not just receipts for completed work. However, income is exclusively money earned that increased the assets of the business. The yearly net income total is the figure that the IRS will use to establish the tax payment.

Income Statement — A report that establishes what the total income is for a certain period of time. This report can be completed by your accountant. It is also known as the profit and loss statement. The financial statement (Figure 14) shows the total financial condition of the business, whereas the income statement shows whether the business made a profit or loss during a particular period.

Indirect Cost — These are items purchased that are not directly identified as a job cost (office supplies, rent, secretary labor, etc.). They may be identified as indirect materials or indirect labor and are considered to be overhead expenses. It's more or less the opposite of direct costs.

Interest Expense — The cost of interest on borrowed money. This is an accounting term but may also be used in estimating practices if interim loans are required to complete a project.

Inventory — All materials held in reserve by the TIB and not identified as belonging to a specific job. This inventory can be considered an asset on the Financial Statement (Figure 14).

Invoice — A document prepared by the TIB to send to the client for payment due on work completed. The same is used by suppliers and subcontractors when billing the TIB.

Itemized Deductions — These are personal expenses incurred by the TIB that are tax deductible. The TIB may file the tax return itemizing these deductions (taxes, interest, medical expenses, charitable contributions, etc.).

Job Cost Accounting — The process of identifying all expenses (such as equipment, material, and labor) on a job in a way so as to capture these amounts on a Job Cost Record (Figure 24). Therefore either the total cost of the job or phases in the job can be evaluated and compared to the original estimate. This is explained in more detail later in the book.

Journal — A record of financial transactions. Each time a transaction is recorded in a journal it is known as a journal entry. There are several types of journal such as the cash receipts journal, cash disbursements journal, and payroll journal.

Liability — The obligation to pay an amount or perform a service. Liabilities can be unpaid invoices, bank notes, taxes due, mortgages, etc. Liabilities offset the assets on the Financial Statement (Figure 14).

Liquid Assets — Cash or assets that can be quickly converted into cash without a reduction in price, such as marketable securities.

Marketable Securities — A near cash asset readily tradeable such as stocks or treasury bills. Securities may be sold with little difficulty because a ready market exists with established market prices.

Net Worth — Also known as owner's equity. The net worth of a business is the value of the business if all bills are paid and assets are liquidated. The remaining cash value is the net worth of the business.

Note Payable — Part of the chart of accounts. It is the account that classifies expenses for regular loan payments, similar to accounts payable.

Note Receivable — The flip side of note payable. If someone owes you money from a loan, the payments are identified as note receivable, similar to accounts receivable.

One-Write System (Figures 20, 21, 22) — A system of cash disbursements and payroll that consists of a set of forms and checks backed with carbon strips. It has a series of holes punched along one end which is attached to a "peg board" used for aligning them. The entries are, therefore, recorded on several documents at one time. Also known as a "peg board system." More information on the One-Write System is discussed later.

Payroll Journal (Figure 21) — A special journal used in recording payroll details including hours worked, gross pay, deductions, and net pay. The payroll journal can be incorporated with the cash disbursements journal using the One-Write System.

Personal Property — Includes all property other than land and buildings (see *Real Property*). Consists of all moveable property, tools and equipment, and accounts receivable.

Pegboard System — See *One-Write System*.

Principal — The original amount of a loan not including the interest or other added expenses.

Profit and Loss Statement — See *Income Statement*.

Real Property — Includes land, land improvements, and buildings. The balance of personal property.

Reconciliation — Adjusting the differences between two items so that the two figures agree. Balancing your checkbook is also known as reconciliation of the bank statement. The accountant will compare the journals with the general ledger and reconcile them so that they balance with one another.

Revenue — The inflow of money to the TIB's business. Total revenue would be the total dollars received by the business. In accounting terms, revenue usually does not include money received from the sale of equipment or borrowed money.

Trial Balance — The trial balance is a tool the accountant uses at the end of each accounting period. It is a worksheet that lists the account balances from the general ledger. Total debits balance with total credits. It simply helps to determine if an error was made in the general ledger entries.

THE ONE-WRITE SYSTEM

The easiest type of recordkeeping/check writing system is called the One-Write System. The TIB will find that this system will save valuable time by combining the check register, cash disbursements journal, and the payroll journal in one easy to use record. The One-Write System (Figures 20, 21, and 22) consists of a pegboard binder, journal sheets, employee earnings sheets, pay statements, and special checks that are prepared with your business name and bank account number. Remember to always keep separate bank accounts for your business and your personal finances. The business records should not include any personal expenses.

Cash Disbursements

Before explaining the how-to of using the One-Write System, it is necessary to set up the check register form as a cash disbursements journal. This procedure will help the TIB to identify and categorize each expense as it is being written. All of the checks written annually by the TIB can be identified as being one of fifteen to twenty types of expense. Before writing any checks a list of these accounts (types of expense) must be established and then written on the check register (as shown on Figure 21). Here is a sample list of regular business expenses:

    Materials
    Equipment Rental
    Subcontractor Labor
    Employee Labor
    FICA/Employment Payment

    FICA/Employer Payment
    Withholding Payment
    Fines and Penalties
    Interest/Bank Charges
    Notes Payable
    Auto Expense
    Utilities
    Insurance
    Professional Services
    Miscellaneous Expenses

This list can be adjusted to meet the particular needs of your type of business. Consulting with your accountant might be helpful when working out this list. As you can see on the check register/cash disbursements journal (Figure 20), there are several columns toward the right side of the sheet. Each one of the account names listed above must be inserted at the top of one of those columns. As checks are written, the amount of the check should be recorded in one of the account columns.

For example, suppose you write Check Number 1586 to White Lumber in the amount of $295.75 for lumber. In the spaces provided on the check, list to whom it is written, the date, check number, and the amount. If the check is written on line 5, then follow line 5 to the column with the heading "Materials." Write the amount in that space. By identifying each check in this fashion, you will easily be able to create reports, complete other journals, post to the general ledger, and gather information for tax purposes.

The check writing procedure holds true for all expenses. If a check is written for a monthly loan payment, write the amount of the check in the column labeled "Notes Payable." If the expense is for repair of a company vehicle, list it under the column labeled "Auto Expense." Accounting and legal fees are identified as "Professional Services."

Miscellaneous Expenses

There are certain expenses that do not fall within the boundaries of any of the categories on the check register/cash disbursements journal. Rather than add extra columns for infrequent expenses, it's best to list these expenses in the column labeled "Miscellaneous Expenses" or "General Expenses." The TIB should take care in describing the miscellaneous expenses so that they can be clearly identified by the accountant. Also, there will be expenses that are not generated by writing a check, such as bank charges. As this expense is generated (usually when you balance the checkbook) just list it in the appropriate column.

The proper procedure is to use separate check register/cash disbursement sheets for each month. As one month ends, simply add up the totals in each column and start a new sheet for the new monthly period. Every month you will know exactly what amount was spent on each category during that month. This is the information the accountant uses to complete the general ledger and the yearly tax return.

Payroll Procedures

When writing a payroll check there are a couple of additional steps. Because there are many one-write systems available, the procedures for completing the forms will vary from company to company. However, as shown on Figures 21 and 22, fold the check register/cash disbursement journal revealing the payroll journal on the back of the sheet. Before writing a payroll check you must first establish the amount. This is done by using the Pay Statement (Figures 21, 22). Record the earnings and pay deductions on the pay statement for each employee. Carbon strips on the back transfer all necessary information to the payroll journal. At the end of each month the totals from each of the columns on the payroll journal will establish the amount of taxes due.

Before actually completing the pay statement take the Employee Earnings Record (Figure 22) and carefully place it (and a sheet of carbon paper) between the pay statement and the payroll journal. Line up the next empty line on the earning record with the appropriate line on the pay statement. This way the pay statement, employee earning record, and payroll journal will all be completed at the same time.

There should be a separate Employee Earnings Record for each employee. Each time a check is written for wages to an employee it should be recorded on his earnings record. With this sheet you will be able to

keep track of the quarterly and yearly earnings on each employee, and you can fill out the quarterly reports and yearly W2 forms much more easily.

During the process of filling out a pay statement, the hours worked, pay scale, FICA, withholding, date of the end of the pay period, and the amount of the check, all are automatically recorded. Unlike other expenses, the payroll information is not extended and transferred to columns on the righthand side of the journal sheet. The record of these expenses is instantly entered into the proper column when the pegboard statement is filled out. These amounts are to be totaled at the end of each month just like the check register/cash disbursements journal.

Now you can write the check. Turn back to the check register and complete it just like any other check. When distributing the employees' pay, enclose the pay statement so that they have a record of their payroll information. Be sure to write each employee's name on each respective statement.

Just Like a Checkbook

Much like a regular checkbook, deposits, bank charges, and bank balances are recorded. Any bank drafts or service charges made by your bank should be recorded and deducted from the balance you show in the journal so that it balances with the bank's monthly statement.

Changing the check register and payroll journal sheets at the end of each month is critical. Everything in accounting is done on a monthly and yearly basis. By making the effort to keep accurate records, and with the help of your accountant, you will be able to establish a functioning system that works for you. Don't let the simple procedures of posting the check amounts into the applicable categories fall by the wayside. It is a must that these amounts be recorded while the memory of what the check was written for is fresh in your mind.

JOB COST

Job costing is a method of recording all of the direct expenses for a particular job and compiling them into an organized report so as to review and evaluate these expenses. The entire success of the TIB's business is founded on "costs." Without an accurate knowledge of costs, a TIB is totally unprepared to engage in the contracting business.

1. It will help when bidding other jobs that may be similar in order to bid them at the right price and still make a profit.
2. It will determine if the work is being done efficiently and economically. If the materials used and labor hours on a particular job are unusually high, the TIB may have an efficiency problem.
3. It will determine whether a profit was made on each job. By comparing the original estimate with the actual costs, the TIB can determine the gross profit (or loss).
4. An accurate job cost will provide the necessary records for preparing a draw or payment request for partially completed work.
5. It will assist in preparing Job Expense Reports (Figure 7) for cost plus contracts. Without a job cost system it will be nearly impossible to insure accuracy in billing.
6. By reviewing the total expenses for a particular job, you will be able to control overpayment of invoices. Unusually high costs on a job will alert a TIB to investigate and may lead to the discovery that an invoice was paid twice or that the supplier overbilled.

The first step in keeping costs on the jobs is by identifying the material and labor expenses on the check register/cash disbursements journal (Figure 20). Use an available column on the register and identify each direct job expense with the name of the job or an identifying number such as Job 101, 102, etc. If a check is written for material or labor on more than one job, break down the expense detail showing the job name for each job and the expense incurred.

Job Cost Payroll

Direct payroll costs as well as material costs must be identified to determine the correct total costs. Just as material, the employees' labor should be recorded for each job. This is easily done through the use of the

Employee Time Card (Figure 23). The time card serves two functions. It records the total hours worked so that the total pay can be determined, and it breaks down the employee hours so that the wages can be allocated to the appropriate jobs. A time card also helps the TIB to remember small jobs that are sometimes forgotten because of the short time it takes to complete them. The time card identifies those jobs for billing to the client. Keep the time card simple. Most construction workers are not naturally detailed in their paperwork abilities.

The goal is to properly record all direct job costs in such a way as to create a Job Cost Record (Figure 24). This record is different from the Job Expense Report (Figure 7) in that it identifies the actual labor costs. The job expense report is used when billing the client on cost plus jobs and has higher labor figures that include overhead expenses. The job cost record is made for the TIB's eyes only. Transfer the information from each check written to the job cost records. There should be a separate record for every job. Each time a check is written it is entered into the check register/cash disbursements journal, identified by job name and check amount, and entered into the appropriate job cost record.

An important note about employees' wages and the job cost record: when dividing the wages into the various jobs, use the gross pay, not the amount of the check. The gross pay includes FICA, withholding, and other deductions. When final calculations are made of the labor costs, a percentage should be added to cover the employer's portions of payroll taxes, insurance, fringe benefits, etc. Those expenses will eventually be incurred and should be included as part of the cost of the job.

Completing a job cost record will be necessary when working on any job. However, job cost systems are only as good as the information provided by the TIB. If the check is recorded accurately and the expense is allocated to the proper account, it will produce accurate and helpful information vital to the success of the business. There are a variety of job cost systems available. As the TIB's business grows, he may want to be more detailed in his identification of the job costs. As many different phases of work (such as carpentry, concrete, drywall, electrical, plumbing, etc.) are entered into, the TIB may eventually generate reports identifying not only the total cost of a job but also the costs of each category of work within each job. This detailed report is very important if the TIB is responsible for several phases of work and several jobs are in progress. Much of this kind of detailed reporting can be done by computer. Refer to your accountant and the section on Computers for more information.

BILLING

When a TIB needs to bill a client for completed work, it is important to provide the client with a written bill, otherwise known as an invoice. The invoice (Figure 25) should be professional in appearance and should contain certain vital information. Your business name, address, and telephone number can be pre-printed at the top of the invoice and can be purchased from most office forms suppliers. Type or write in the client's name and address in the space provided. The invoice can be either typed or written, as long as it is written very clearly and neatly.

The invoice must include enough information so the client fully understands what services were rendered for the amount being billed. Include information such as the job name and address, the phase of the job, type of work, and any references to contract documents that might be helpful. One way to show the financial information is to state first the total contract price, then add change orders (if any), show the total of the previous invoices, then show the balance left to draw on the job (as per contract agreement).

Finally, subtract the amount that is due now which leaves the balance left to be billed (as per contract agreement). Using this format will make it easy for both you and the client to keep track of the billing progress. Try to use helpful information such as percentages, dates, identifying numbers, etc., on each invoice so that everyone involved can understand and process the invoice quickly and efficiently. If necessary, a detailed report of the work completed (relating to this invoice) could be attached if the client needed additional verification.

Monthly Statement

For some trades who do several jobs for one client each month, it may be necessary to send out a statement (Figure 26) at the end of each month to clients who were issued invoices during that month. After each job the client should be sent an invoice. The invoices should be sent no later than the end of the month. The statement is a less detailed bill which summarizes the month's activities. It lists the invoices by date, number, and amount. Many TIBs allow the client to pay only after the monthly statement is received. The statement insures that the client hasn't lost one of the invoices. However, if a TIB does not complete several jobs for one client during the month, the statement may not be necessary.

Cash Flow

Cash flow is very important to any business. Too often an inefficient billing system delays the flow of cash, hindering the TIB's ability to pay the bills. It is obvious that the sooner a TIB bills his clients, the sooner he will get paid. The longer the billing procedures are delayed, the longer it will take to get the payment.

Accounts Receivable

In conjunction with the billing function, you should keep an Accounts Receivable List (Figure 30). Every time an invoice is sent out, the amount of the invoice, date, and client's name is recorded on this list. The Accounts Receivable List will keep a record of what is due and when it is due. When the invoice is paid, payment should be noted on the list. As invoices become overdue, the TIB can follow up quickly because all of the billing information has been compiled onto one document.

Without a clear record of what has been billed and paid, a TIB could easily lose track of an overdue invoice. If you rely on the clients to notify you of an inability to pay, you're going to be waiting a long time. It's your money, so keep track of it.

Overdue Invoices

When a bill is due, be very careful how a client is approached about payment. It is important to "read" a client before addressing the issue. If there was a problem

ACCOUNTS RECEIVABLE JOURNAL

Page __1_ of _1__

INVOICE DATE	DESCRIPTION	AMOUNT	DATE DUE	DATE REC'D
xxxx	GBJ Builders (129 E. Peach)	450.00		
xxxx	Jim Gray (Home Remodel)	1943.00		
xxxx	Bolder Construction (Residence)	2500.00		
xxxx	J & K Construction (BFK Bldg.)	1500.00		
xxxx	B. Jones (Jones Bldg.)	750.00		
xxxx	Wall Enterprises (City Bldg.)	2396.50		
xxxx	Barker Company (Smith Remodel)	565.80		
xxxx	J. Keller (Billings Factory)	4878.00		
	TOTALS			

FIGURE 30

during the course of the job, he might be holding back payment. Many times a kind word and a little understanding of his point of view will work wonders. If he is a client you want to do more work for, the worst thing to do is to irritate him by harassing him about payments.

If it becomes obvious that you will not be paid for work completed, it may be necessary to stop all work on his projects until past due bills are made current. It's not something anyone wants to do, but it may be an absolute necessity. Letting overdue clients go too long without payment may eventually result in the demise of your business. Refer to the information on lien rights to determine whether filing a lien is a wise course of action.

An easy yet professional way to handle overdue clients is to send out a series of notices. To start with, send the client a friendly reminder (Figure 27). This first letter should be sent no later than two to three weeks after the payment is due. If this does not generate a response within two to four weeks after the first letter was sent, send a more serious letter (Figure 28). If the client has not issued payment within a reasonable time, the TIB may need to take serious action by sending the client a demand letter (Figure 29). This procedure is acceptable, but must be weighed on the basis of the TIB's relationship with the client.

Cash Receipts Journal

Billing and the accounts receivable ledger have to do with receiving money. As this money is received, it must be identified by more than just a date or check on the accounts receivable ledger. Every check or money received must be described so that the reason for the income can be documented. The Cash Receipts Journal (Figure 31) is an excellent way to document these transactions. As soon as a client pays a bill, money is refunded for materials, or interest is paid by a bank account, it must be entered in the Cash Receipts Journal. This is also a good way to cross reference with accounts receivable to verify whether a client has paid an invoice.

PAYING THE BILLS

Bills will come in by the ton, and keeping track of how much is due can be a monumental job. This task can be simplified by keeping an Accounts Payable Ledger (Figure 32). This list will help keep track of what is due, when it is due, and when it has been paid. It will keep you informed of when and how much is the total amount needed for these bills.

When a bill is received, immediately enter it into this ledger and enter the date it was received. When each invoice is paid, mark the date paid. Also, before filing away each invoice with all supporting documents attached, mark it paid with a notation of the date and check number.

You may want to keep your unpaid invoices in a file folder that can be purchased at any office supply store. This folder simply allows you to file, alphabetically, all invoices or statements as you receive them. When you pay each invoice or statement, after writing the check number and date paid directly on the invoice, file it away in the permanent file for that vendor.

You should have individual file folders for each of your vendors (suppliers and subcontractors). This will enable you to have quick access to any information regarding materials purchased from various suppliers or costs of subcontract labor for each vendor.

Discounts Will Save You Money

Some suppliers will give a discount on their invoices if they are paid within a certain period of time. These discounts usually range between 1 and 2%. This may not seem like much money, but over a year's time it can add up to a substantial amount. Be very attentive to the discount due dates on the accounts payable ledger. This will be your main record and reminder of when to pay the invoices. You should strive whenever possible to take advantage of each and every discount available.

On the other hand, by paying invoices past the due date you may be subject to late charges. Like discounts, these charges can really add up. Take the late charges and add them to the discounts that could have been obtained by paying on time, and you can quickly see the remarkable amount of money that is being wasted every year by so many TIBs. Learning discipline and business management skills that help you to pay your bills on time will produce a substantial increase in annual business income.

ACCOUNTS PAYABLE JOURNAL

Page __1__ of __1__

DATE REC'D	ACCOUNT	PROJECT	AMOUNT	DISCOUNT DUE DATE	DATE PAID
xxxx	Williams Hardware	BFK Bldg.	762.45		
xxxx	City Power Company	Utility	23.00		
xxxx	Lands Electric Service Company	Billings Fac.	456.00		
xxxx	Parker Office Supply	Office	12.76		
xxxx	First State Insurance	Insurance	175.00		
xxxx	White Lumber Company	BFK Bldg.	223.90		
xxxx	Talon Paint	BFK Bldg.	30.00		
xxxx	Harvest Supply	BFK Bldg.	59.00		
xxxx	Drywall Supply Co.	Jones Bldg.	112.88		
xxxx	City Steel	Jones Bldg.	98.00		
xxxx	Central Equipment Rental	Jones Res.	100.00		
xxxx	Miller Accounting Service	Prof. Serv.	75.00		

FIGURE 32

Taxes

Most TIBs do not understand and do not make provision for the various taxes that every business must pay. That is why so many businesses no longer exist. Perhaps these are harsh words, but the reality is that every year many TIBs close their businesses because they were unaware of the tax requirements and of the penalties for not paying the taxes on time. Many businesses close within the first five years for this reason. However, by the fifth year, most TIBs have found out, one way or another, what the various levels of government expect regarding taxes. Unfortunately, many are currently in serious debt as a result of being unprepared for the heavy tax burden.

The problem begins when the TIB, rather than setting the tax money aside, spends it on bills, equipment purchases, or personal expenses. "Oh, I would never do that," many of you might say. Oh, yes you would — mainly because you may not realize that the money spent was not yours to spend. Not realizing this fact is usually due to lack of proper records showing exactly how much money must be set aside for taxes, and when this money should be sent to the government.

The guidelines in this book are procedural instructions only. The tax structure may change every year, and the TIB should consult his accountant for investment and tax guidance.

Let's examine each type of tax and explain its function.

PAYROLL TAXES

This is probably the biggest area of trouble for most TIBs. Primarily because when issuing a payroll check, the employee gets only the net amount of his pay. The FICA and withholding are noted on the check, but the money is still in the TIB's bank account and is included in the bank balance. It becomes too easy to spend money without regard to the responsibility of later paying the money to the government.

On top of what is withheld from the employee's paycheck, the government will also require the TIB to pay an additional percentage of the employee's gross pay as part of the employer's participation. Identified below is each portion of the payroll taxes.

W-4

To know how much tax to withhold from employees' wages, you should have a form W4 on file for each employee. Every employee, when starting work, should fill out and sign this form. Copies of the W-4 can be obtained from your accountant or from the local IRS office. You are required to ask each employee for a new W-4 by February 15 of each year. Even though the employee completes and signs this form, you are required to verify the Social Security number from this official government-issued card.

Withholding

When writing a payroll check, one of the items taken out of the gross pay is the income tax withholding. The TIB should first ask the IRS for a copy of the Employer's Tax Guide (Circular E). The Tax Guide changes every year so make sure you have a current copy. All of the information concerning payroll taxes and withholding is included in this guide.

To use the charts shown in the Tax Guide, first determine the employee's marital status. Second, identify how often the employees are paid — monthly, bi-monthly, or weekly. Using this information you then find the correct tax chart for that employee. Locate his gross wages under the wages column and cross reference this amount with the number of withholding allowances claimed on his W-4 (0 through 10). The amount listed is that employee's withholding for that pay period.

FICA (Social Security)

In addition to income tax withholding there is also the FICA withholding (Social Security). This amount can also be figured using the Employer's Tax Guide. Use the chart labeled Social Security Employee Tax Table, which bases the FICA on the employee's gross pay. The procedure for figuring the FICA is similar to that for withholding.

After the withholding and the FICA amounts are figured, they are subtracted from the employee's gross pay. Shown in the record-keeping procedures outlined in this book, the amounts withheld from an employee's pay are routinely recorded into the payroll journal when the check is written. By adding the amounts in the payroll journal for both the FICA and withholding, you can determine what amount is owed to the government.

Employer Extra FICA

In addition to the FICA withheld from the employee's check, the IRS requires the employer to match a certain amount of this tax. For instance, in addition to the 7.51% (this amount is subject to change each year) that is paid into Social Security by the employee, the employer must pay an additional 7.51% of the gross pay. This extra portion is paid to the IRS with each tax payment. Refer to the Circular E publication each year for changes in the tax structure and ask your accountant for up-to-date information. This extra amount paid by the employer is hard to remember and can get you into lots of trouble if you forget. When setting the wages for new employees, remember this percentage of additional wages.

How to Pay Payroll Taxes

As previously stated, the IRS must be notified so that they can assign any new business a nine-digit number called an Employer Identification Number. Without this number, the tax payments will not be correctly identified by the IRS. If you have not asked for a number, the IRS will send you the Form SS-4 so that you can apply.

The IRS will then send you deposit slips (Form 8109) with your new identification number printed on them. Use these deposit slips each time you pay the payroll taxes. Take a check for the taxes and a deposit slip to your bank, and they will process it for you. Be sure to "black in" the appropriate boxes showing the type of tax and the tax period dates that the deposit covers. For example, the payroll taxes should be designated as Form 941 taxes.

When to Pay Payroll Taxes

The sum total of both the FICA (both employer and employee portions) and the income tax withholding must be paid quarterly. The payment due dates for each quarter are January 31, April 30, July 31, and October 31. Late payments can result in heavy penalties. There are times, however, when the FICA and withholding taxes must be paid more frequently than every three months.

The amount of taxes withheld will determine how often to make these deposits. If, for example, your taxes are $500.00 or more but less than $3,000.00 at the end of any month, you must make your tax deposit within fifteen days after the end of that month.

If the taxes are $3,000.00 or more at the end of any eighth-monthly deposit period (each month is divided into eight deposit periods that end on the 3rd, 7th, 11th, 15th, 19th, 22nd, 25th, and last days of the month), you must make your deposit within three banking days after the end of that eighth-monthly deposit period. All the rules and examples of when to make these deposits are stated in the Employer's Tax Guide. Read it carefully to get a full understanding of when you must make these deposits.

Remember, these rules may change from year to year, and every employer must keep informed of the changes.

FUTA

One tax that will slip up on you without warning is the Federal Unemployment Tax. This tax is paid quarterly. To determine whether you must deposit tax for any of the first three quarters in a year, multiply the employee's total annual gross wages that were paid during the quarter by .008.

If the amount is less than $100.00, you can wait and deposit it next quarter with the tax figured for that quarter. But if it is over $100.00, you must deposit it within one month after the end of the quarter. The FUTA tax is applicable to the first $7,000.00 of an employee's annual gross wages only. You will use the same Form 8109 coupon deposit slips that you did for FICA and withholding, but be sure to "black in" the box showing that this is a Form 940 tax. Also, show to which quarter the deposit is to be applied.

W-2

At the end of every calendar year, you must give a Form W-2 to each employee from whom you withheld income tax. This Form W-2 must also be sent to any employee whose wages were subject to Social Security taxes. These forms must be mailed by January 31 of the year following that in which the tax was withheld. Therefore, it is very important to keep separate payroll tax records for each employee throughout the year, so that it doesn't take too long to figure the year-end tax information. Again, more information concerning the W-2 can be found in the Circular E Employer's Tax Guide.

STATE TAXES

Contact your local tax office for information on the employer tax requirements for your state. Hopefully, your accountant is knowledgeable regarding the applicable state and local tax laws.

Obviously, it is important to keep good records and to pay the government the correct amounts and on time. Don't let complacency or a "don't care" attitude keep you from fulfilling the duties of an employer.

PROPERTY TAXES

There are taxes that property owners must prepare for each year, such as property and school taxes. If you or your business own vehicles, equipment, land, buildings, etc., the locality, state, county, and/or city will send you a bill based on the appraised value of the property.

If you buy land for investment purposes, remember that if you own it on December 31, you will have to pay the taxes for the year. Taxes are paid on property during the closing process when selling the property. These taxes are usually figured on a prorated basis by the closing agent. If you sell and close on a piece of property, be sure that the taxes are figured correctly. Always keep a copy of the closing statement in case you have to prove how much you paid to the applicable government tax office. If you sell the property, go to the tax office and notify them in person that you sold the property, and give them the name of the individual who is now the owner. Otherwise, they may not get their proper identification and could continue to send you an annual bill for the taxes.

Each applicable tax office will send you a bill at the end of each year, based on the appraised value of the property. If you do not agree with the appraised value, there is usually a standard protest procedure. In many cases, you can get the appraised value lowered.

EMPLOYER'S PERSONAL TAXES

As an employer, you are not above withholding taxes on your personal paycheck. To operate a business properly, the TIB must consider himself an employee of the business and should receive a set wage (smaller than you would like), regularly deducting withholding and FICA.

When you are self-employed, you are taxed at a higher Social Security rate. This is known as the self-employment tax. This tax has been the doom for many TIBs who were unprepared for the heavy payment at tax time. The self-employment tax rate can be as much as 14% depending on the tax law changes each year. If a TIB doesn't consistently deduct from his wages to cover the self-employment and withholding taxes, he may have an uncomfortable meeting with the banker around April of each year, trying to get a loan to pay his taxes. Don't take shortcuts. Do it right.

TAX DEDUCTIONS

Tax evasion can lead to an extended stay in prison; tax avoidance, however, is the right of every taxpayer. To avoid paying unnecessary taxes, you must consistently keep a record of the business expenses that are allowable business tax deductions. Here is a list of some that are common expenses. Contact your accountant for details on these and other allowable deductions.

Accounting Fees — The costs of financial advice, bookkeeping service, tax preparation fees, financial statements, and the cost of an IRS audit.

Advertising — Any cost of advertising and promoting the business. This includes newspaper ads, telephone book listings, and brochures.

Charitable Contributions — Donations to qualified charities and churches. If the business is a corporation, the donations are business deductions. If a TIB is not incorporated, they are a deduction, but not a business expense.

Child Care — At least part of the costs for child care is deductible, if it is essential for you to be able to work.

Commissions — Fees to agents or managers are deductible.

Dues and Subscriptions — The membership costs of professional or business organizations, subscriptions to business-related magazines or journals are all deductible.

Education — You may deduct courses, seminars, books, tuition, and travel and lodging expenses to maintain and improve your business skills.

Entertainment — Meals at restaurants or at your home to entertain business associates and clients are valid business expenses. However, the IRS tends to scrutinize this deduction, so it is best to keep a record of: 1) the purpose of the gathering, 2) who attended, and 3) each person's business relationship to you.

Gifts — Gifts to clients or customers are deductible as well as gifts to people who work for you. However, there is a $25.00 deductible limit for clients and a $1,600.00 total limit for your employees. This may change as the tax laws change.

Home Office — The IRS has placed strict guidelines for home office deductions. To qualify, the space used for the office must be your principal place of business, set apart from the rest of the home. It must be used on a regular basis and for business purposes only. Personal activities cannot be done in the office space. The percentage of the space used in the home is the deductible percentage of all of the utility and mortgage expenses on the home.

Insurance Premiums — Fire, theft, storage, liability, employee health, workman's compensation, and other types of insurance are business deductions.

Interest Payments — Interest payments on business loans are deductible. Finance charges on credit cards are deductible, if considered a business expense.

Legal Fees — Legal fees for business purposes are deductible.

Office Equipment — The costs of office equipment used for business purposes such as desks, calculators, typewriters, answering machines, computers, etc., are all deductible.

Office Supplies — You may deduct the cost of office supplies used in the business such as paper, pens, pencils, stationery, business cards, file folders, envelopes, etc.

Operating Losses — Payments due from clients that were never collected are deductible.

Postage — All stamps and mailing expenses for the business are deductible.

Professional Services — The cost of professional services used in the course of operating the business are deductible. These include accounting, advertising, typing, answering service, cleaning, etc.

Repairs and Maintenance — Repairs to equipment such as typewriters, calculators, computers, etc., are deductible.

State and Local Taxes — When these taxes are directly connected with your business, they are deductible as a business expense.

Telephone — Long distance business calls are deductible.

Travel—Expenses related to travel for business purposes are deductible such as meals, lodging, tips, baggage checks, etc. Travel by car can be deducted two ways: the operating expenses for the car (gas, oil, parking, etc.) or a mileage deduction. Your accountant can advise as to which is better for you.

Transportation—If your office is in the home, transportation to the jobs, bank, suppliers, etc., is deductible. If the primary place of business is not in the home, the cost of driving to the office from the home is not deductible.

ACCOUNTANTS

Certified Public Accountants, like lawyers, are typically quite expensive. Their services should be used on a limited basis. However, without the help and advice of a CPA, you might find yourself in serious financial trouble, so use their services as much as you need.

Every CPA has certain areas of expertise. Some are not experienced in dealing with problems of certain businesses. Many specialize in retail, some in construction, and some in personal tax returns. Their specialty or expertise mainly depends on their past clients and years of experience, so be sure your CPA has experience working with people in your type of business.

Many businesses hire accountants when they actually need bookkeepers. Although accountants know how to do bookkeeping, they are not bookkeepers. On the other hand, some TIBs hire a tax preparer when they really need an accountant. Their primary function is to analyze your books and prepare tax returns based on the figures in them.

A good accountant will be a management consultant, helping you to understand you total financial picture and make wise decisions in all areas of your business. For instance, he will give you advice on whether to purchase equipment, hire employees, or incorporate your business. If you need a business loan, an accountant can prepare the necessary financial reports. He can help you plan strategic financial moves that will save taxes each year. He will also represent you should the IRS audit your tax return.

There are many CPAs who aren't very knowledgeable about the new tax laws, and as a result are so timidly conservative that you end up paying more taxes than necessary. This is not to say that they should cheat or be dishonest, but that it takes an aggressive and skilled individual to dig out and find every available deduction. Experience will teach an aggressive and bright CPA how to honestly save the client big bucks.

In an accounting firm of any size, you may deal directly with the head man, but most of the work may be done by his assistants. This procedure may be just fine with you, but be certain that you are receiving the attention and service you desire.

Find a CPA you feel comfortable with and who can talk intelligently with you about your type of work. An experienced CPA can provide valuable information and assistance that will affect every aspect of the business.

The future goal of every TIB should be, with the accountant's help, to establish a record keeping system that is 90% controlled and operated by the TIB. The accountant will then receive certain reports, generated by this record keeping system each month or each quarter for tax purposes. It will take time for a TIB to operate without the accountant's frequent help, but eventually these procedures will become much easier.

11. LEGAL

ATTORNEYS

The usefulness of a good attorney should not be minimized because from time to time throughout the course of normal business transactions, legal services will be required.

Because of the expense, however, the use of legal advice should be controlled. An attorney will charge for letters, phone calls, meetings, stamps, lunches, and anything else for which he spends time or money on your behalf. Be direct and concise when meeting with your attorney.

Legal help won't be necessary when starting the business. Getting a tax number and filing the assumed name of the company can be accomplished with little or no assistance. It is a good idea, however, to let an attorney take care of the process of forming a corporation or partnership.

Attorneys can advise you on other legal matters such as real estate contracts, lien rights, construction contracts, and other legally binding documents. The average TIB will require the services of an attorney only on occasion, and usually only to review a legal document to determine its effect on your business.

There is no need to call your attorney every time someone threatens to sue you. Those are idle threats more often than not. However, don't hesitate to find out your legal position on a matter if the situation becomes serious. Attorneys are for your protection. Going to court should be the very last alternative. A skilled lawyer can usually settle a matter long before it reaches the court.

LIEN RIGHTS

Although there are several types of liens, we will discuss only one important type called the mechanic's lien. Basically, this type of lien is filed by a TIB or supplier's attorney after they completed work on a project but have not been paid for that work by the client. Its purpose is to guarantee that the TIB will be paid for work done for the client.

In today's market, the mechanic's lien is a necessary and viable tool. If you have not been paid for work completed, the filing of a lien may be absolutely necessary to produce the desired payment.

The lien won't necessarily force the client to pay what is owed, but it will do two important things. First, it will legally tie up a client's project so that if he wants to sell the property, the lien will have to be settled before he can complete his transaction. Second, if the client tries to obtain financing for his project, the lien will hold up his loan until it is resolved. In many ways the lien has more of a psychological effect than anything else.

An important thing to remember is that there is a limited time that a TIB is legally allowed to file a lien. Seek a lawyer's advice to find out the limits. Never let a customer go past ninety days overdue without taking some sort of action. You don't want to make enemies, but too many individuals will take advantage if you don't stand up for your rights. The attorney can take care of all the paperwork, which makes this process very easy. Clients understand the necessity of liens and shouldn't be alarmed or surprised if you are forced to file. Sometimes the simple fact of filing produces payment without ever going to court.

CONTRACTUAL AGREEMENTS

Much has already been said about contracts. It is worth reaffirming the need for legal help from time to time on this very important issue. Contracts can be very confusing for the inexperienced. Because every contract is different, it takes time to understand and, if necessary, write a contract.

If a client offers a contract for you to sign, and you are not sure as to what is says or what your legal obligations will be, then let your attorney take a look at it. The client will understand that extra time is needed for this review. Never sign anything that you don't fully understand. Make this a firm rule and you will never regret it.

Quite a bit of work, especially in the construction industry, is done with only a verbal agreement. This has certainly worked from time to time, but the risks are extremely high. No matter what kind of relationship you have with the client, always, without fail, have the agreement in writing signed and dated by both parties.

Without a contract the client has no legal obligation to pay you a dime. Of course, there is a moral and ethical obligation, but when it comes to money changing hands, the worst side of a person's character is often exposed.

In addition, so many misunderstandings have arisen when two individuals have only a verbal contract or agreement. Because we assume everyone else thinks like we do, there doesn't seem to be a need for further clarification of the agreement. Unfortunately, this is not the case, and costly mistakes are the result. Avoid simple misunderstandings. Protect yourself with a written contract.

12. INSURANCE

WORKMEN'S COMPENSATION

Until about 1910, if an employee suffered a work-related injury or contracted an occupational disease, he had to prove that the employer was negligent before he could obtain any compensation. Eventually it was decided that accidents were bound to occur and regardless of who was at fault, insurance should be provided to cover these injuries. Today, almost every state requires employers to carry workmen's compensation insurance. Even in the states where it is not the law, any business having at least one part-time employee is at risk of liability for an accident if the employee can prove negligence on the part of the employer.

The premiums are set by the government and are based on the total estimated payroll and on each employee's job classification. In order for the insurance company to establish the required monthly payments for this policy, they must examine each employee's responsibility and determine his or her job classification. The more dangerous the job, the higher the rate. Therefore, make sure that each employee's job classification and rate is correct so that you don't pay high rates for non-dangerous positions. The insurance company will make adjustments if the changes are valid.

Because it is impossible to estimate accurately an employer's payroll for the next six to twelve months, the premiums are estimated based on the estimated yearly payroll. These may be either high or low, depending on the actual payroll for the year. To determine the exact amount due, the insurance company will conduct a yearly audit of the payroll records. This audit will determine the actual payroll expenses for each employee and, based on the job classification, will establish the exact amount of premium due for the required coverage. If there was an overpayment by the employer, the insurance company will issue a refund. However, the TIB must pay extra if the audit shows that the monthly payments were too low.

If your monthly payroll is very large, it might be best to complete your own monthly labor report. A form provided by the insurance company can be filled out and sent in every month so that the monthly payments are fairly accurate. This procedure is relatively easy, if the payroll records are kept organized. When the yearly audit is made by the insurance company, the TIB will not be faced with any big surprises of large payments due.

The workmen's compensation coverage is for the employee's benefit, but is effective only during the hours he is on the payroll. If injured on the job, this policy will cover the medical expenses for those injuries and will pay a portion of the employee's salary during the time he is unable to work. This policy does not cover injuries that occur away from the job. In general, the policy provides compensation to the employee for 1) medical and surgical care, 2) fatal injuries (pays the last illness expenses and makes a weekly payment to the dependents for a specified period), and 3) temporary total disability (the employee receives a percentage of his weekly wage at the time of his injury up to a specified amount). Usually the same rate applies when he is permanently disabled.

Subcontractors
If a TIB uses subcontract labor, it should be company

policy to use only subs who carry a workmen's compensation insurance policy. Have each subcontractor's insurance company send a copy of the Certificate of Insurance verifying that the sub is complying with the law. Keep each of these Certificates of Insurance in your files, because when the insurance company audits your company at the end of the year, you will be charged extra for insuring the subcontractors who did not carry a policy during the time they worked on your jobs.

If you hire a subcontractor who does not carry a workmen's compensation policy, it should be standard practice to deduct from his check the amount that the insurance company will charge extra to insure him while working on your job. Likewise, if you are a subcontractor and do not carry workmen's compensation insurance, the contractor for whom you are doing work may agree to deduct from your contract price enough money to cover the premiums he will be charged by his insurance carrier. Most contractors prefer, however, that you provide your own coverage. Not carrying this insurance may cause you to lose work that you might otherwise have gotten.

GENERAL LIABILITY

Unlike workmen's compensation, the general liability policy is not a legal requirement. However, a TIB takes a tremendous risk by not carrying some type of liability insurance. Particularly in the construction industry, damage to property and personal injury can easily results in tens of thousands of dollars in expenses.

The coverage of a general liability policy applies to bodily injury and property damage that is caused by you or someone connected with your company in connection with work performed by your company. Injury or property damage caused by an employee during his off hours in most cases is not a company responsibility. For this reason, a TIB should keep close tabs on each piece of equipment and the location and duty of every employee at all times.

The rates for general liability insurance vary depending on the type and amount of work and the amount of coverage. Obviously, the more dangerous the work

and the larger the jobs, the higher the insurance rates.

Insurance rates go up every year, and we all wish we could do without it or find an alternative. Unfortunately, there isn't much of an alternative for the small TIB. However, all insurance companies are not the same, and they don't have the same rates or the same coverage. Shop around until you find the company that is right for you.

Some insurance companies will try to sell you a policy with fabulous and unnecessary coverage. Get only what you need based on your type of work and your gross annual sales. Ask other TIBs about their coverage to get an idea of the liability policy you need.

AUTOMOBILE INSURANCE

All vehicles owned by you personally or by your company should have some form of insurance. In most states it is mandatory for every vehicle owner to carry a minimum coverage of liability insurance. Whether it is required or not, you should always have at least a liability policy particularly when, as the vehicle owner, you are responsible for the actions and safety of an employee driving that vehicle.

If any of your employees will, at any time, drive a company vehicle, the driver must be individually named on the policy. If an employee has a wreck in your vehicle and is not named on the policy, the insurance company may not (probably won't) pay for the damage.

An auto liability policy covers the bodily injury and property damage to other persons or their property caused by you or one of the covered employee drivers. The rates will be based on the driving records of each of the named drivers. If you have an employee with a terrible driving record, it is wise to omit his name from the list of insured drivers in order to lower the rates. Then instruct the employee never to drive one of your vehicles.

A collision policy is the flip side of liability. Collision covers damage to your vehicle or personal injury to you or your passengers. Collision rates vary depending on the year and make of the vehicle and amount of coverage requested. Don't get more coverage than the vehicle is worth. If you get a loan at the bank to buy a vehicle, the bank will require you to carry collision insurance. Remember this expense when deciding whether you can afford to purchase a new vehicle.

It's also a good idea to have what is known as an "uninsured motorist" policy. This type of coverage is very inexpensive and pays for your vehicle's damage or the passengers' bodily injury that was the fault of a motorist who did not carry liability insurance.

As high as insurance rates are, it's a good idea to restrict the privilege of driving your vehicles to only the individuals who absolutely have to. Also, check to see if any of your many insurance policies have a duplication of coverage. For instance, if you have automobile liability then your general liability policy shouldn't overlap and provide coverage for the same items.

BONDING

Generally, jobs that will require bonding should be avoided by the newly established TIB. In order to obtain such bonds, you must have been in business for a number of years, and you will be required to furnish the bonding company with financial statements, profit and loss statements, and lists of previous jobs completed for a period of three years or more.

The cost of these bonds is based on the total dollar amount of the contract for the job requiring the bond. The bond is good for that job only, and is not a blanket insurance coverage as in the case of liability insurance.

Bonds are required for most government work unless a TIB works as a subcontractor to a bonded general contractor whose bond covers the TIB's portion of the job. The government or client will establish the bonding limits for each job. Bond requirements for a particular job are usually an indication that there are many other strict and time-consuming "red tape" procedures involved.

New and inexperienced TIBs should avoid jobs that require bonds. However, as a track record is

developed and the opportunities arise, this type of work may be part of the TIB's future.

The three most common types of bonds are the Bid Bond, the Performance Bond, and the Payment Bond. Although these bonds are not available to the newly established TIB through normal channels, the Small Business Administration has a program that may be of assistance to the new contractor who is having trouble obtaining bonding. This program is authorized to guarantee up to 90% of losses incurred under bid, payment, or performance bonds on contracts up to $500,000.00. You might want to check with the Small Business Administration office in your area for more information and applications. A simple definition of each type of bond follows.

Bid Bond
The bid bond is a guarantee that the bid given to the client is good, and that the TIB will enter into contract with the client if he is awarded that job for the amount of his proposal. Backing out of a bid that is bonded will hinder the TIB's ability to get future bonds.

Performance Bond
The performance bond is a guarantee to the client that the TIB will complete the work as prescribed by the plans and specifications and the terms of the contract. The charge for performance bonds runs from threequarters of one percent to one percent of the full amount of the contract. Backing out of a performance bond can cost big bucks depending on the amount of a contract.

Payment Bond
The payment bond is a guarantee to anyone dealing with the bonded contractor that they will be paid for their services or goods. This guarantees the client that, in the event that a TIB is dishonest or in financial trouble, the materials and labor supplied for the job will be as specified and completed, and paid for as originally agreed. It gives him confidence that the TIB's suppliers or subcontractors can't successfully execute a mechanic's lien as a result of an unpaid bill.

These are very oversimplified definitions. Your insurance company or bonding company can provide more in-depth information as the need arises.

SECTION V
SUMMING IT UP

13. HOME ENVIRONMENT

FAMILY IS PRIORITY ONE

It doesn't seem right to place such an important and vital topic toward the end of this book. Without a healthy family atmosphere the stress of the business can become unbearable. Every person, whether a business owner or not, needs the love and support of the family unit.

Too many businessmen in this country have devoted so much time to building up their business that their families are being neglected. More time is spent developing business relationships than is spent developing family relationships. Your spouse and your children need your regular attention.

Business owners fail to see that the strain of the business doesn't just affect the boss but also every member of the family. Taking time to communicate with your spouse about the business, and explaining the details that are questioned will go a long way toward helping her or him in understanding and coming to terms with the situation, thus becoming needed support.

It's true that money can't make you happy. But, more specifically, the love of money is the root of the prob-lem. The love of money makes people willing to sacrifice life and family to gain it. In this selfish quest for a better life, are you willing to sacrifice the very thing that will bring happiness? Peace in the home is something that you cannot put a price tag on.

If you want to run your own business, do it with your priorities in order. Work hard and yes, make sacri-fices, but don't forget that necessary time with the family. Sacrificing may mean less TV time or sports activities or whatever you do in your spare time. It is very important for every TIB to exercise time man-agement. In other words, manage your time wisely making the very best use of your time every hour of the day.

Remember, when it's time for family, resist the pres-sure to allow business to interrupt. There is nothing that can replace the love and support of your spouse and your children. To achieve this support will re-quire your building a strong bond between you and your family.

A stable home environment will produce a brighter outlook on your day-to-day business activities, and it will improve your ability to deal with problems and to handle stress. Home is where it all starts.

14. BACK TO THE FUTURE

PLANNING FOR THE FUTURE

So much of what happens in small businesses today is a result of circumstance, rather than planning and preparation. Few TIBs realize that they should have very definite goals for the future and a plan for how to reach those goals. The future cannot be predetermined and circumstances cannot be manipulated, but you don't have to be subject to the circumstances ruling your decisionmaking process.

In other words, every phase of your business can be planned to some degree. Hiring employees, buying equipment, office and warehouse requirements, and diversification are all areas that can and should receive careful thought toward future goals.

Every six months you should sit down with your spouse and discuss every phase of the business. Get away from the house, away from the kids, and away from the phone, and give this discussion all the uninterrupted time it deserves. Take out a whole day, if necessary, to discuss nothing but future goals and plans. Write out your goals for the next six months. Then write out the goals for the next year and then for the next five years.

Be very specific about your plans. Leaving an area of the business out of your planning will cause that area to suffer for lack of direction.

Planning Personal Affairs

Planning your personal affairs is just as important as planning for the business. If one suffers, the other will suffer. You and your spouse should also write a plan for your personal affairs, just as you did for the business. Discuss areas such as vacations, children's education, future homes, savings, investments, financial budgeting, etc.

It's hard to make yourself sit down and plan for the future. But unless you plan, all of these areas will be subject to an "on-the-spot" decision that may or may not be what your business or personal affairs really need.

BRANCHING INTO OTHER FIELDS

In the early years, there is so much going on, so many new areas to learn, that you really don't have time to think about expanding or diversifying. However, every company will grow and so many companies, particularly in an upward economy, would rather grow quickly than to set the slow pace necessary to insure a healthy growth. Sometimes it's hard to resist the pressure to grow fast, and many TIBs give in to that pressure only to find themselves and their companies out of control.

The first years of any business must be dedicated to developing the business management skills. Those business skills must have time to catch up with the trade skills that put you into business. It takes time to learn how to run a business properly.

Eventually, if you follow the principles stated in this book, you will master business techniques and management skills. At that point you may be prepared to branch out into other areas. This is not to imply that you should do anything your heart desires, but every TIB has more than one skill that can be slowly and carefully developed to a level of expertise, just as you did with your primary trade.

Serve a Larger Market

Your business, like all businesses, is functioning and making money selling your services to a particular group of buyers, which we will call your market. You are providing a certain service to meet a certain demand for that service. This is what economists call "supply and demand." The market demand for any industry, including construction and your trade, will not continue to have the same demand. The economy and the demands for your services may drastically change from year to year. No one, not even the economists, can accurately predict what will happen.

By expanding your services you can tap into a larger market which will give you a larger chance of survival if the demand for your other services diminishes.

Most TIBs have far greater talents and skills than they realize. Developing those skills must be done at a careful and steady pace to guarantee strong growth. Be sure not to expand too quickly with a skill in which you cannot provide the same quality and service which your clients have come to expect.

Keep your eyes and ears open, always thinking about the future. You don't have to fabricate new fields of endeavor. The opportunities are there, and when the time comes be ready to make your move.

WHEN TO THROW IN THE TOWEL

This is a tough subject. So much has been said about how to stay in business and how to make your business grow. It is necessary, however, to deal with an issue that so many TIBs have had to face. There is definitely time when a businessman must close or cut back his business.

You might think that it would be obvious when a business needs to be cut back. Not necessarily. When it becomes obvious it is usually much too late for productive correction, resulting many times in severe financial trouble and possible bankruptcy. There are symptoms that give indications of serious problems. Recognizing these symptoms can prevent a lot of anguish.

Businesses that do not have an accurately functioning recordkeeping system will not know whether or not they are making money. Some may begin to notice the need to get the payments from the clients more quickly each month in order to cover expenses. As this trend continues, it will, ever so subtly, put them into financial jeopardy. This is not to say that everyone who experiences cash flow problems is in serious trouble. However, without an operating recordkeeping system, there is very little opportunity to get an accurate picture of your situation.

What happens is that as materials are purchased and subcontractors hired, their bills may not be sent for thirty to forty-five days after the work is completed, depending on their billing system and the time of month the materials and work were ordered. In the meantime, as your need for cash increases, a partial or complete payment is received from a client. This money is not spent on bills for his job but on overdue bills on other jobs. When the bills for his job come in the money is not available because it was spent on bills for previous jobs and to cover payroll. Then the money needed for these new bills has to come from the proceeds of the next job. As the distance between the job's being completed and paid for and the payment of the bills for that job increases, there grows a need for more jobs to pay the ever increasing amount of outstanding bills. This is simply a case of robbing Peter to pay Paul.

If you are not making a large enough profit on your work at least to carry the overhead, then it will be only a matter of time until you are in serious financial trouble. This money shifting can only be kept up until the work flow slows down or stops temporarily. Then, even though there is no incoming money, there continues to be a steady flow of bills, bills, bills.

One of the first things to do when this type of situation arises is to cut the overhead as much as possible. Cut it and cut it quickly. The sooner the overhead expenses are lowered, the better the chance of survival. If you have an office or warehouse space you rent monthly, if must be eliminated. Employees who are not absolutely necessary must be laid off immediately. Return rental equipment that is not being used, and reduce your salary to only what you must have to support yourself and your family.

If steps are taken soon enough, you may be able to salvage the situation and stay out of serious debt. Many people get very badly burned when bankruptcy is filed. Don't even consider this as an option. Make good on your commitments. One of those commitments was when you told the suppliers that if they sold you materials on credit, you would pay them for those materials. They have bills to pay just like everyone else, and they have trusted you to keep your word.

If you find yourself unable to pay bills on time, it is a good idea to contact personally every creditor before he contacts you. Explain the situation and the steps you are taking to resolve it. Be honest and sincere. They deserve and will appreciate your full attention and concern for their position. Remember, it's their money, not yours.

An important thing to remember is that while some businesses have a positive net worth, most of that worth may not be in liquid assets. Their financial worth is tied up in assets that are not easily turned into cash. Even if the appraised value of these assets is justified, a rush liquidation in order to get working capital can produce bargain basement prices. Being "cash broke" can eventually mean really broke if you sell your assets too quickly. This is why advance planning is a must so you will have the needed working capital available.

It may be appropriate to get counselling and advice from another businessman who is sympathetic to your situation and who has the wisdom and insight to help you bring it under control. It's a difficult road to travel. With the right guidance and hard work, you may be able to overcome the problems and eventually become debt free.

Sometimes, even with all the effort and dedication a TIB can muster, a time may come when all that is left is to "throw in the towel." With as much honesty and integrity as possible, clean up the wreckage and prepare yourself to reenter the world as a tradesman. You have not totally failed; you are still a competent, skilled craftsman. There will be better days ahead and you now have the advantage of knowing where the problems arose should you decide to enter the business world again. When you try again you will be much more knowledgeable and skilled as a business manager, and you will have the advantage of knowing in advance many of the answers to the problems of becoming a Tradesman in Business. Failure is not final.

15. CONCLUSION

In today's marketplace there are thousands of TIBs. Most of them are skilled technicians. They have worked for years developing their tradesman abilities. The key to their success is the realization that they have only half the skills necessary to operate a business. No matter how small a business is, it will take management and organizational skills to keep it running. First comes the technician, then comes the business manager. A TIB's success will be determined by his ability to make this critical transition.

The many different topics discussed in this book may be overwhelming for a new TIB. It may seem impossible for you to learn and function expertly in each phase of the business. Don't abandon the information just because adapting these skills to your business seems too difficult. Remember, if you are committed to being self employed, you are committed to operating a business, and you will face the same problems day after day with or without the proper management skills. Wouldn't it be much easier if you at least tried to put these principles into practice? Take it one step at a time and one day at a time. It's not as difficult as you might think.

As you grow as a manager and as your business grows, take advantage of the multitude of resources that will help you in business development. There are several sources that a TIB can tap for useful information in many areas of business: your accountant and attorney, peers and associates, and printed material and seminars. Each one of these provides a link in the business management chain; without one of those links the chain may prove to have a weak area.

Certainly the accountant and attorney can provide professional advice and counseling that will help keep a TIB out of hot water. Professional advice is seldom a waste and usually a benefit. Peers and associates have all experienced a variety of business situations with both success and failure. Their experience will be similar to yours and they will provide practical advice. Find those individuals who are trustworthy and wise, and allow their experience to teach you better management skills.

Hopefully this book has been a challenge as well as a source of information. A challenge not only in becoming a better businessman, but also in expanding your creative boundaries. We challenge every TIB to stretch his or her skills, abilities, and creativity to improve their self confidence and to operate their businesses with honesty and integrity. We challenge you to challenge yourself, to raise your standards of quality in every aspect of business, and never to be satisfied with a job that does not meet those standards.

There are rules to learn and dues to pay. You can make it with hard work and a little business smarts. You can succeed. You don't have to be a statistic; you can be a winner. Think positively at all times, but keep a firm grip on reality. Face your problems head on. We believe in Tradesmen in Business. We also believe that with the proper training and guidance the trend of business failures can be drastically turned around. With the help of committed, hard-working TIBs, it will be.

SECTION VI
RESOURCE MATERIALS

APPENDIX

Index of Publications 135

Accounting
Blueprint Reading
Building Codes
Business Guides for Contractors
Carpentry
Computers
Concrete and Formwork
Contractual Agreements and Construction Forms
Definitions of Construction Terms
Drywall
Electrical
Estimating and Bidding
Excavation and Heavy Equipment
General Building and Construction Guides
Government Resources
Health and Safety
Heating, Ventilating, Air Conditioning
Insulation
Legal
Masonry
Office Supplies and Business Forms
Painting and Wall Covering
Plastering
Plumbing
Remodeling
Roofing and Sheetmetal
Steel Construction
Surveying
Welding

Resource Addresses 153

Bibliography 158

Forms **159**

Inventory Sheet
Purchasing Worksheet
Purchase Order
Work Schedule
Areas of Code Influence
Bid Worksheet
Job Expense Report
Subcontractor/Supplier Bid Sheet
Telephone Bid
Sample Proposal Cover Letter
Estimate
Proposal
Change Order
Sample Financial Statement
Application for Draw
Business Resume
Application for Employment
Letter of Appreciation
Letter for Client's Employee
One-Write Check-Writing Systems
Time Card
Job Cost Record
Invoice
Statement
Reminder About Overdue Payment
Appeal Letter on an Overdue Payment
Demand Letter on an Overdue Payment
Accounts Receivable Journal
Cash Receipts Journal
Accounts Payable Journal

ACCOUNTING

Basic Accounting for the Small Business
Self-Counsel Press, Inc.
A down-to-earth manual on how to be a better book-keeper and understand simple accounting practices for a business. Concisely written and easy to read.

Bookkeeping Made Simple
Doubleday & Company, Inc.
A practical, easy-to-use guide for people in business who want to keep their own books.

Builder's Guide to Accounting Revised
Craftsman Book Co.
Step-by-step, easy to follow guidelines for setting up and maintaining an efficient record keeping system. It shows how to meet state and federal accounting requirements, and explains what the Tax Reform Act of 1986 can mean to your business.

Dictionary of Accounting
The MIT Press
Easy to understand definitions of over 1300 accounting terms.

How to Learn Basic Bookkeeping in Ten Easy Lessons
Barnes & Noble Books
A simple workbook giving easy to understand instructions with sample problems to work out. For business owners who want to learn simple bookkeeping.

Modern Construction Accounting Methods and Controls
Prentice-Hall, Inc.
Offers a model accounting system suitable for nearly all contractors. Includes charts of accounts, journals, job cards, daily and weekly reports, etc. It also has forms and procedures for recording labor and related payroll taxes, forms for allocating and controlling overhead, forms and procedures for recording cash disbursements, etc.

Walker's Practical Accounting and Cost Keeping for Contractors
Frank R. Walker Co.
A detailed and informative guide for contractors. Covers bookkeeping systems, methods for keeping time, labor and material, equipment, and payroll. Also shows how to prepare profitable estimates, draw up contracts, and keep records required for operating a construction company.

BLUEPRINT READING

Answers on Blueprint Reading
Audel/Macmillan Publishing Co.
A complete instruction manual for understanding blueprints of machines, electrical systems, and architecture. A question and answer format plus clear illustrations.

Blueprint Reading for the Building Trades
Craftsman Book Co.
How to read and understand construction documents, blueprints, and schedules. Includes layouts of structural, mechanical, and electrical drawings, how to interpret sectional views, and how to follow diagrams.

Electrical Blueprint Reading
Craftsman Book Co.
Shows how to read and interpret electrical drawings, wiring diagrams, and specifications for construction of electrical systems in buildings.

National Electrical Code Blueprint Reading
American Technical Publishers, Inc.
This fully illustrated workbook teaches print reading while acquainting readers with applicable sections of the 1987 Electrical Code.

Print Reading For Welders
American Technical Publishers, Inc.
This graphic workbook depicts areas of importance to welders, making it easier to interpret welding prints. Explains welding and shop terminology and symbols.

Reading Construction Drawings
McGraw-Hill Book Co.
Dealing with every area from vocabulary to site plans, this is a step-by-step evaluation of how to properly and accurately read construction drawings.

Residential Print Reading
American Technical Publishers, Inc.
This heavily illustrated workbook presents print reading from securing a permit through a study of final trim.

BUILDING CODES

Contractor's Guide to the Building Code
Craftsman Book Co.
Explains in plain English exactly what the uniform building code requires and shows how to design and construct residential and light commercial buildings that will pass inspection the first time. Suggests how to work with the inspector to minimize construction costs, what common building shortcuts are likely to be cited, and where exceptions are granted.

Model Energy Code
Council of American Building Officials

National Electrical Code
National Fire Protection Association

National Building Codes
Building Official & Code Administrators International

One and Two Family Dwelling Code
Council of American Building Officials

Standard Building Codes
Southern Building Code Congress International, Inc.

Understanding Building Codes and Standards in the United States
National Association of Home Builders
Explains how the code process works and how code organizations conduct their business.

Uniform Building Code
International Conference of Building Officials

BUSINESS GUIDES FOR CONTRACTORS

AGC Guide to Compliance with the Immigration Reform and Control Act of 1986
The Associated General Contractors of America
Includes line-by-line instructions on how to complete the new forms, questions and answers on the new requirements, and a copy of the law itself.

Bonding for Subcontractors
Explains what a bond is and describes the kind of information needed to get the proper surety bond. For every subcontractor, this manual provides basic "must know" information.

Complete Secretary's Handbook
Prentice-Hall, Inc.
Provides solutions to every on-the-job problem from accounting to zip codes.

Construction & Maintenance Daily Log
Safety Meeting Outlines
A hardbound log used by contractors, superintendents, and foremen for keeping accurate daily job records.

Contractor's Guide to Change Orders
Prentice-Hall, Inc.
Explains how to win top payment for potential delays and hidden expenses. Helps you find price and get paid for contract changes.

Contractor's Survival Manual
Craftsman Book Co.
How to survive hard times in construction and take full advantage of the profit cycles. "Real life" solutions to problems such as what to do when the bills can't be paid, finding money and buying time, transferring debt, the alternatives to bankruptcy, lawsuits, judgments, and liens.

Contractor's Year Round Tax Guide
Craftsman Book Co.
How to set up and run your construction business to

minimize taxes. Covers tax shelters, writeoffs, and investments that will reduce your taxes, accounting methods, and what the IRS allows and what it often questions.

Construction Operations Manual of Policies and Procedures
Prentice-Hall, Inc.
Clearly spells out all your business procedures — both in the office and out in the field. All you have to do is insert in the blank spaces your company's name or the name of the person responsible for each task — and the manual is ready to use. Covers project records and controls, personnel administration, and much more.

"Daytimer" Daily Planner
DayTimers, Inc.
One of the recommended daily planners and appointment books available in most office supply stores.

"Dayrunner" Daily Planner
Harper House, Inc.
One of the recommended daily planners and appointment books. Available in most office supply stores.

Handbook of Business Letters
Prentice-Hall, Inc.
A quick and easy guide to writing business letters with 761 sample letters covering every possible business situation. Each is written with a style and tone that's perfectly suited to any occasion.

Hiring and Firing
National Association of Plumbing-Heating-Cooling Contractors
How to legally and fairly hire and terminate employees. It covers all the bases on employee/employer relations from a legal standpoint. Easy to use guidelines are provided on how to keep records on your employees. Information on federal and state laws governing employment practices is also included.

Homemade Money
Betterway Publications, Inc.
Practical, expert advice on starting and running a home-based business. Includes A to Z section on business basics.

How an S Corporation Can Save You Tax
Enterprise Publishing, Inc.
A leading tax accountant shows how an "S" corporation can save you literally thousands of tax dollars. Learn how to set up an S corporation, how it affects investments, tax shelters, when not to set up an S corporation, and much more.

How to Subcontract and Supervise Hired Personnel
National Association of Home Builders
Manage for results and cut the cost of directing workers and subcontractors. Follow the steps to implementing a productive management program and get the most for your labor dollar.

Inside the Family Business
National Association of Home Builders
This practical guide to the do's and don'ts of family business deals with the dynamics of family relationships and how they affect the business.

Insurance for Subcontractors
The American Subcontractors Association, Inc.
Provides basic understanding of the various insurance policies normally purchased by subcontractors, from general liability and workmen's compensation to builder's risk insurance and equipment floaters.

Job Descriptions for the Construction Industry
PAS, Inc.
117 actual construction job descriptions ready for use. Presented in standardized, easy-to-use format. Construction job descriptions range from owner to receptionist.

Model Personnel Handbook for Subcontractors
The American Subcontractors Association, Inc.
A general collection of model personnel policies written especially for subcontracting firms. Once the sample policies are adapted to your company's needs, the handbook can help orient your new employees, increase worker productivity and morale, and minimize the potential for management liability.

Overhead Manual Series

The American Subcontractors Association, Inc.
Track and recover the hidden overhead expense on every job with a three-part manual series. These workbooks explain overhead recovery and management, give sample figures, and provide procedures for assigning expenses.

Productivity in the Residential Building Trades

National Association of Home Builders
Evaluates how production crews can improve their output. Presents case studies of productive time for the various trades. Analyzes causes of nonproductive time.

Straight Talk About Small Business

McGraw-Hill Book Co.
A realistic and candid guide for the person considering starting his or her own business. It explains the hard work, sacrifices, and risks that go with a small business as well as the profits and satisfactions.

Summary of State Regulations and Taxes Affecting General Contractors

American Insurance Association
Provides information on licensing and tax responsibilities for contractors in each state.

Supervisor's Factomatic

Prentice-Hall, Inc.
Tells you exactly what to say, what to do, or what to write for the everyday people problems. It tells you how to explain to a valuable employee that the big raise you hinted at won't fit in the budget; what to say to an employee accused of sexual harassment; how to tell your employees to get to work on time without hurting morale. Instant answers to over 500 of your toughest people problems.

The Construction Manager

Prentice-Hall, Inc.
Control the countless details that are a part of every construction job. A daily planner and appointment desk book designed exclusively for construction professionals.

The McGraw-Hill Construction Business Handbook

McGraw-Hill Book Co.
Topics of vital importance to contractors, subcontractors, and material suppliers in the construction industry. Covers organization, accounting, taxes and record keeping, finance, insurance and bonding, contract analysis, federal government regulations, contract performance, contract rights, collection procedures, and financial difficulties.

CARPENTRY

Basic Carpentry Techniques

Ortho
All the techniques you need for everything from building shelves to framing a house.

Carpenters and Building Library

Audel/Macmillan Publishing Co.
This comprehensive four volume library includes Volume I: tools, steel square, joinery; Volume II: builder's math, plans, specifications; Volume III: layouts, foundations, framing; and Volume IV: millwork, power tools, painting.

Carpentry for Residential Construction

Craftsman Book Co.
How to do professional quality carpentry work in homes and apartments. Both rough and finish carpentry are covered in detail; planning the job, the estimated man hours required, and a step-by-step guide to completing each part.

Carpentry in Commercial Construction

Craftsman Book Co.
Covers forming, framing, exteriors, interior finish, and cabinet installation in commercial buildings.

Carpentry Layout

Craftsman Book Co.
Explains the easy way to figure: stairs, rafters, joists, studs, and pickets. Shows how to set foundation corner stakes. Practical examples show how to use a handheld calculator as a powerful layout tool.

Completed Siding Handbook
Audel/Macmillan Publishing Co.
Comprehensive step-by-step instructions on every aspect of siding a building.

Finish Carpentry Techniques
Ortho
This guide tells do-it-yourselfers everything they need to know to finish a framed building.

Finish Carpentry
Craftsman Book Co.
The time-saving methods and proven shortcuts you need to do first class finish work on any job. Includes details for all aspects of finish carpentry.

Guide to Residential Carpentry
Glencoe Publishing Company
Everything is covered step by step from design and planning to foundations, framing, roofing, interior and exterior finishing.

Handbook of Wood Technology and House Construction
Research and Education Association
A comprehensive source in building complete houses, adding to existing houses, and home improvement projects. A useful reference with numerous illustrations for the practical designer and builder.

Lightweight Steel Framing Systems Manual
Metal Lath/Steel Framing Association

Roof Framing
Craftsman Book Co.
Frame any type of roof in common use today — even if you've never framed a roof before. Over 400 illustrations take you through every measurement and every cut on each type of roof.

The Complete Foundation and Floor Framing Book
TAB Books, Inc.

The Pocket Size Carpenter's Helper
National Association of Home Builders
A wealth of essential information on residential construction. Designed for on-the-job use, has simplified span charts, nailing schedules, stair rules, and calculating loads for beams and posts, and much more.

Wood-Frame House Construction
Craftsman Book Co.
From the layout of the outer walls, excavation and formwork, to finish carpentry and painting, every step of construction is covered in detail, with clear illustrations and explanations.

COMPUTERS

Computers: The Builder's New Tool
Craftsman Book Co.
Shows how to find the right computer system for your needs. Takes you step by step through each important decision, from selecting the software to getting your equipment set up and operating.

Construction Computer Applications Directory
Construction Industry Press
A complete listing of computer software for the construction industry.

Contractor's Computer System Specifier
Construction Industry Press
Forms used in analyzing your business computer needs in order to determine what kind of software you need.

Contractor's Computer Systems Evaluator
Construction Industry Press
Forms that evaluate the available software to help you choose which package is best suited to your business.

NAHB Software Review Program
National Association of Home Builders
After reviewing all construction software, the NAHB has chosen a list of software that best meets the needs of the building industry.

Software Catalog for Home Builders
National Association of Home Builders
Contains detailed descriptions of computer software programs available to the residential construction industry.

CONCRETE AND FORMWORK

Cast-In-Place Walls
American Concrete Institute

Concrete and Concrete Construction
American Concrete Institute
Provides concrete contractors insight into the problems and solutions associated with high quality concrete construction.

Concrete and Formwork
Craftsman Book Co.
This practical manual has all the information you need to select and pour the right mix for the job, lay out the structure, choose the right form materials, design and build the forms, and finish and cure the concrete.

Concrete Construction and Estimating
Craftsman Book Co.
Explains how to estimate the quantity of labor and materials needed, plan the job, erect forms, handle the concrete into place, etc.

Concrete Construction Handbook
McGraw-Hill Book Co.
Materials for concrete, properties of concrete, formwork and shoring, finishing and curing, precast and prestress concrete, and much more.

Concrete Formwork
American Technical Publishers, Inc.
Detailed, heavily illustrated information on all aspects of formwork, from site preparation through concrete placement and form stripping.

Connections for Tilt-Up Wall Construction
Portland Cement Association
Connection elements for tilt-up wall construction.

Construction for Drilled Pier Foundations
John Wiley and Sons, Inc.
This handbook details all phases of the construction of drilled pier foundations. This is the most up-to-date and comprehensive book on its subject available.

Form Builders Manual
National Association of Home Builders
This manual presents techniques, designs, examples, and related information on form building.

Formwork for Concrete
McGraw-Hill Book Co.

Handbook of Concrete Technology and Masonry Construction
Research and Education Association
Preferred economical construction procedures are presented with extensive illustrations to provide practical know-how in form designs, reinforced concrete, concrete blocks, mortar, stone masonry, and brick and tile masonry.

Recommended Practice for Precast Concrete
Prestressed Concrete Institute
Establishes basic guidelines which should be considered by the precast concrete manufacturer and the precast concrete erector.

Residential Concrete
National Association of Home Builders
Covers basic concrete practices and how to recognize most faulty procedures. Answers questions about concrete scalling and spalling.

Slabs on Grade
American Concrete Institute
This book gives helpful information on the basic properties of concrete and what it takes to make good concrete. Concrete fundamentals are outlined. Good construction practices are described.

Standard Practice for Curing Concrete
American Concrete Institute
Basic principles of curing are stated; commonly accepted methods, procedures, and materials are de-

scribed. Requirements are given for curing pavements and other slabs on ground; for structures and buildings; and for mass concrete.

CONTRACTUAL AGREEMENTS AND CONSTRUCTION FORMS

AGC Manual of Contract Documents
The Associated General Contractors of America
A library of contract documents for contractors. This is a complete collection of the available AGC contract forms and instructions. Each of the forms can be ordered separately by requesting their free catalog.

AIA Catalog of Contract Documents
American Institute of Architects
An extensive listing of contract forms for contractors and architects. Contact the AIA for their free contract documents price list.

CSI Catalog
Construction Specifications Institute
A listing of contract forms for the construction industry.

Means Forms for Construction Professionals
R.S. Means Company, Inc.
Over 100 full size prototype forms and worksheets in a three-ring binder. For all construction activity — estimating, designing, project administration, scheduling, etc. Each section is preceded by an instruction which explains how to use the forms most effectively.

ASA's Standardized Forms
The American Subcontractor's Association, Inc.
ASA has standard subcontractor forms available such as work authorization form, application for payment form, contract change order form, project information form, etc.

Contract Documents
National Electrical Contractors Association
A reference book published by Associated Specialty Contractors, Inc. containing educational materials on what subcontract agreements should and should not contain. Copies of AIA subcontract forms with instructions, procedures for change order, punch lists, damages, payments, overtime, etc.

NEBS Business Forms for Contractors
New England Business Systems
Free catalog of available forms includes subcontract agreements, proposals, and change orders.

Plans, Specs and Contracts for Building Professions
R.S. Means Co.
Answers virtually any question about construction documents. Aids you in making a step-by-step review of any special document you're working on — bidding package, project manual, bonding or insurance, contract general conditions. It helps you sidestep dozens of major and minor conflicts typically met by contractors.

Walker's Contractors Office Supply Catalog
Frank R. Walker Company
Free catalog of forms for contractors. All types of estimating, bidding, and accounting forms available. Includes time sheets, job cost forms, subcontract agreements, etc.

Winning the Battle of Subcontract Forms
The American Subcontractors Association, Inc.
Explains how to avoid costly errors by reading and understanding contracts before you sign. Offers alternatives for making your subcontracts with the general contractors as neutral and fair as possible.

DEFINITIONS OF CONSTRUCTION TERMS

Architectural Building Trades Dictionary
American Technical Publishers, Inc.
Hundreds of helpful illustrations and more than 7500 architectural and building trade terms. Includes a glossary of important legal terms related to building trades and a complete listing of common material sizes.

Construction Glossary: An Encyclopedia Reference Manual
John Wiley & Sons, Inc.

Dictionary of Architecture and Construction
McGraw-Hill Book Co.
An extensive listing of architecture and construction terms.

Encyclopedia of Building and Construction Terms
Prentice-Hall, Inc.
Definitions of over 2,400 "must know" terms. Includes a helpful index which categorizes each term by trade.

Illustrated Dictionary of Building
John Wiley & Sons, Inc.

Illustrated Dictionary of Building Materials and Techniques
TAB Books, Inc.
Defines and explains construction terms used by electricians, carpenters, masons, plumbers, painters, HVAC, and architects.

The Means Illustrated Construction Dictionary
R.S. Means Company, Inc.
More than 12,000 definitions commonly used in construction work. Covers every area of construction and architecture.

DRYWALL

Drywall Contracting
Craftsman Book Co.
How to do quality drywall work, how to plan and estimate each job, and how to start and keep your drywall business thriving. Covers the eight essential steps in making any drywall estimate.

Drywall: Installation and Applications
American Technical Publishers, Inc.
Step-by-step procedures with sequential drawings cover the basic principles of all types of drywall construction.

Gypsum Construction Handbook
United States Gypsum Company
A unique reference, workbook, applications manual, product catalog, student text, and specifications guide. The industry standard for gypsum construction.

ELECTRICAL

Basic Wiring Techniques
Ortho
A complete guide to wiring in the home. Illustrated with step-by-step procedures.

Electrical Construction Wiring
American Technical Publishers, Inc.
Electrical theory and wiring procedures are covered with hands-on activities and numerous detailed illustrations.

Guide to the 1984 Electrical Code
Audel/Macmillan Publishing Co.
This is the most authoritative guide available to interpreting the National Electrical Code. It clarifies the code through examples and illustrations. All terms and regulations are fully explained.

Handbook of Modern Electrical Wiring
Craftsman Book Co.
Covers planning, installing, and testing most electrical work. Formulas and tables are included so you can calculate resistance, impedance, branch circuit loads, and power factors.

Housewiring
Audel/Macmillan Publishing Co.
The rules and regulations of the National Electrical Code as they apply to residential wiring are fully explained with detailed examples and illustrations.

Industrial and Commercial Wiring
American Technical Publishers, Inc.
Contains special requirements for industrial and commercial wiring. How to install and service electrical equipment of all types from motors to swimming pool fixtures.

Journeyman Electrician's Workbook
American Technical Publishers Inc.
Provides important information for electricians seeking to upgrade job skills. Helps prepare for the Journeyman Electrician's Exam.

Master Electrician's Workbook
American Technical Publishers, Inc.
Designed for the journeyman electrician preparing for the Master's Exam.

Manual of Electrical Contracting
Craftsman Book Co.
From the tools you need for installing electrical work in new construction and remodeling to developing the finances and skills you need to run your business.

NECA Standard Installation
The National Electrical Contractors Association
A reference and guide to good workmanship and electrical construction containing tables of dimensions for raceway spacing and the spacing of supports.

ESTIMATING AND BIDDING

Carpentry Estimating
Craftsman Book Co.
Simple and clear instructions on how to take off quantities and figure costs for all rough and finish carpentry. Provides checklists, extensive man-hour tables, and worksheets with calculation factors built in that take the guess work out of carpentry estimating.

Construction Estimates from Take Off to Bid
McGraw-Hill Book Co.

Cost Records for Construction Estimating
Craftsman Book Co.
How to organize and use cost information from jobs just completed to make more accurate estimates in the future.

Electrical Construction Cost Estimating
McGraw-Hill Book Co.
Detailed explanation of electrical cost estimating, covering all phases of electrical construction.

Estimating and Analysis for Commercial Renovation
R.S. Means Company, Inc.
Detailed methods for inspection and evaluation of existing structures. How to choose the right estimating technique. Includes forms, charts, checklists, and pricing sheets.

Estimating Control Systems for HVAC
McGraw-Hill Book Co.
This very practical book is designed for HVAC contractors who need guidelines for estimating commercial temperature control systems.

Estimating Earthwork Quantities
Norseman Publishing Co.
A unique how-to estimating system that simplifies earthwork estimating.

Estimating Electrical Construction
Craftsman Book Co.
A practical approach to estimating materials and labor for residential and commercial electrical construction. Explains how to use labor units, the plan takeoff, and bid summary to establish an accurate estimate.

Estimating Plumbing Costs
Craftsman Book Co.
Basic procedures for estimating materials, labor, and direct and indirect costs for residential and commercial plumbing jobs. Explains how to interpret plot plans, meet code requirements, and make accurate takeoffs.

How to Estimate Building Losses and Construction Costs
Prentice-Hall, Inc.
Gives all the time-saving procedures, formulas, and up-to-date guidelines and tables needed to prepare or check building damage estimates. Covers every insurable type of damage.

Means Electrical Estimating

R.S. Means Company, Inc.
Using a unique estimating "module" technique, it quickly directs you to the appropriate electrical estimating procedure. Each module examines an installation segment in detail with a cutaway illustration.

Means Interior Estimating

R.S. Means Company, Inc.
Examines all phases of construction for interior estimating. It has dozens of easy to follow prototype estimating forms and diagrams of interior assemblies, graphics, and drawings.

Means Landscape Estimating

R.S. Means Company, Inc.
Covers the site visit, plan review process, quantity takeoff, material, labor and equipment pricing, overhead, and profit calculations, proposals, and much more.

Means Mechanical Estimating

R.S. Means Company, Inc.
Assists in making a thorough review of plans and suggests take-off procedures, includes forms, examples, and steps for pre-bid scheduling for labor and equipment usage.

Means Structural Steel Estimating

R.S. Means Company, Inc.
How steel fabrication plants work; shows how to evaluate the bid package and make the structural steel takeoff; explains steel erection techniques and estimating for miscellaneous iron and metals.

Means Unit Price Estimating

R.S. Means Company, Inc.
Helps you to prepare better unit cost estimate. Describes the most productive, universally accepted ways to estimate. Has checklists and charts with procedures to insure nothing is forgotten in your estimate. A model estimate for a multi-story office building is included.

PDCA's Estimating Guide

Painting and Decorating Contractors of America
A guide to all facets of estimating, containing information on labor production and material usage for various operations performed by painting and wallcovering contractors. Covers residential, light commercial, and volume work.

Process and Industrial Pipe Estimating

Craftsman Book Co.
A clear, concise guide to estimating costs of fabricating and installing underground and above ground piping.

Remodeling Estimates Reference Book

National Association of Home Builders
A comprehensive estimating guide for the professional remodeler. Emphasizes products, material costs, and labor costs.

Rule of Thumb Cost Estimating for Building Mechanical Systems

McGraw-Hill Book Co.
Aids the user in accurately estimating system costs on the basis of as little as preliminary floor plans of a building and a few tentative elevation drawings.

Structural Concrete Cost Estimating

McGraw-Hill Book Co.
Guide for concrete and formwork estimating.

Walker's Building Estimator's Reference Book

Frank R. Walker Company
Includes new construction methods and types of material available. Quantity takeoffs and material and labor costs. An excellent reference for contractors.

Walker's Remodeling Estimator's Reference Book

Frank R. Walker Company
Covers interior and exterior residential remodeling. Develops precise estimates for each phase of the job — kitchens, baths, additions, stairways, plumbing, electrical, etc.

Wallcovering Estimating Guide

Painting and Decorating Contractors of America
A handy pocket-size reference designed to improve skills and knowledge required to accurately estimate wallcovering. It also includes charts and tables to simplify calculations.

EXCAVATION AND HEAVY EQUIPMENT

Excavation and Grading Handbook, Revised
Craftsman Book Co.
Explains how to handle all excavation, grading, compaction, paving, and pipeline work, setting cut and fill stakes, working in rock, unsuitable material or mud, compaction tests, trenching around utility lines, and much more.

Operating the Tractor-Loader-Backhoe
Craftsman Book Co.
Explains how to get maximum productivity from this highly versatile machine. Each task is illustrated with diagrams and photographs.

Pipe and Excavation Contracting
Craftsman Book Co.
How to read plans and compute quantities for both trench and surface excavation, figure crew and equipment productivity rates, estimate unit costs, bid the work, and get the bonds you need. Covers equipment required, asphalt and rock removal, and how to avoid costly pitfalls.

GENERAL BUILDING AND CONSTRUCTION GUIDES

Architectural Graphics Standards
John Wiley & Sons, Inc.
Prepared by the American Institute of Architects, it provides vital information for designers of all types. Considered an industry standard.

Builders and Contractor's Guide to New Methods and Materials in Home Construction
Prentice-Hall, Inc.
From laying the foundation to selecting roofing materials, this unique, easy-to-use resource shows you how to save money at every stage of a home construction project.

Building Construction Illustrated
Van Nostrand Reinhold
A manual of residential and light construction containing step-by-step guidelines with over 1000 illustrations.

Complete Building Construction
Audel/Macmillan Publishing Co.
A comprehensive guide to constructing a frame or brick building. This fully illustrated volume provides all the necessary information from laying out the building to finishing the inside.

Construction Scheduling Simplified
Prentice-Hall, Inc.
A practical, how-to guide to scheduling a construction project.

Construction Techniques — Volumes 1 and 2
National Association of Home Builders
A valuable resource for reliable, first-hand information on the methods, tools, and materials needed for quality home construction.

Handbook of Architectural Details for Commercial Buildings
McGraw-Hill Book Co.
Fully illustrated examples of architectural details for all types of commercial construction.

Homeowner's Encyclopedia of House Construction
McGraw-Hill Book Co.
Thorough and detailed explanations of each phase of residential construction.

House Construction Details
McGraw-Hill Book Co.
Offers the most current hands-on information available on the design and building of single family dwellings. An excellent guide to time-saving tips, use of factory made components, and insights into new materials and designs.

Means Graphic Construction Standards
R.S. Means Company, Inc.
This fully illustrated guide helps you to analyze and select building assemblies. Accompanying charts provide man hours per unit. Based on proven construction methods.

GOVERNMENT RESOURCES

Department of Labor, local office. For information relating to your status as an employer or independent contractor contact your local Labor Department, Wage & Hour Division. Ask for a copy of "The Fair Labor Standards Act of 1938 as amended."

Department of Labor, Washington, DC offers two free booklets entitled "More Than A Dream: Raising the Money and Running Your Own Business," and "Employment Relationship Under the Fair Labor Standards Act." If you plan to hire help for your business, direct inquiry to the Wage & Hour Division.

Internal Revenue Service. Among the many free publications of interest to small business owners are:
Tax Guide for Small Business, No. 334
Business Use of Your Home, No. 587
Index to Tax Publications, No. 900
Depreciation, No. 534
Tax Withholding and Declaration of Estimated Tax, No. 505
Determining Whether a Worker is an Employee, No. SS-8
Self Employment Tax, No. 533

Small Business Administration (SBA). For a complete listing of the SBA publications, request SBA 115A (Free SBA Publications) and SBA 115B (For Sale Booklets). Among the free booklets available are "Business Loans from the SBA" and "Incorporating a Small Business."

HEALTH AND SAFETY

Guide for Voluntary Compliance with OSHA
The Associated General Contractors of America
A "what to do" aid for contractors in coping with the U.S. Department of Labor's "Safety and Health Regulations for Construction."

Guide to Drug Free Job Sites
The Associated General Contractors of America
A manual to assist contractors in dealing with on the job drug and alcohol problems. This comprehensive guide represents the opinions of experts on questions and issues to be addressed when considering the development of programs to eliminate substance abuse problems on the job. The guide covers management options, legal aspects, and drug program elements as well as sample company policies, notices, forms, and supervisory guidelines.

Manual of Accident Prevention in Construction
The Associated General Contractors of America
This manual is devoted to recommended safe practices on construction work. Each section deals with a different phase of construction, yet they all apply to any type of work.

Safety Program Assistance for Subcontractors
The American Subcontractors Association, Inc.
Offers practical guidelines for creating and managing a safety program for any small to medium-sized company. Outlines, step by step, the basic elements necessary to get your safety program up and running; also provides samples of the forms needed to maintain accurate safety records.

HEATING, VENTILATING, AND AIR CONDITIONING

Accepted Industry Practice for Industrial Duct Construction
Sheet Metal and Air Conditioning Contractor's National Association
This guide is a compilation of standards which have wide acceptance in industry for the fabrication and installation of ducts.

Air-Conditioning: Home and Commercial
Audel/Macmillan Publishing Co.
A complete guide to the construction, installation, operation, maintenance, and repair of home, commercial, and industrial air conditioning systems.

ARI Standards
Air-Conditioning and Refrigeration Institute
ARI has over fifty booklets available to the HVAC contractor concerning every aspect of mechanical construction.

Heating, Ventilating and Air Conditioning Library
Audel/Macmillan Publishing Co.
A three-volume reference library for those who install, operate, maintain, and repair HVAC equipment commercially, industrially, or at home. Each volume is illustrated with photographs, drawings, tables, and charts.

HVAC Contracting
Craftsman Book Co.
Your guide to setting up and running a successful HVAC contracting company. Shows how to plan and design all types of system for maximum efficiency and lowest costs, and explains how to sell your customers on the designs you propose. Describes the right way to use all the instruments, equipment, and reference materials essential to HVAC contracting.

HVAC Systems — Duct Design
Sheet Metal & Air Conditioning Contractors National Association
A duct system design manual for commercial and light industrial heating, ventilating, and air conditioning systems which has been structured to offer the designer options in design methods, materials, and construction.

Residential Duct Systems
National Association of Home Builders
Aids in developing plans for heating, ventilation, and air conditioning systems to avoid oversizing and keep your customer's energy bills lower. Also useful for retrofitting HVAC systems.

Job Supervisor's Manual
Mechanical Contractors Association of America, Inc.
Comprehensive, specific, and straightforward, this manual covers the entire spectrum of job management from job site organization to punch lists.

INSULATION

Safety Handbook for the Insulation Crafts
National Insulation Contractors Association
A pocket-size handbook for distribution to all insulation personnel. It is a brief and practical guide to help reduce and eliminate job hazards. It contains an employee sign-off form for documentation use in your company safety program.

The Superinsulated Home Book
John Wiley & Sons, Inc.
Gives details for assembling the construction elements of the superinsulated house. Includes 250 plates and architectural drawings.

Walker's Insulation Techniques and Estimating Handbook
Frank R. Walker Company
A detailed but not highly technical reference book covering today's insulation products, how and where they are used, their costs, and the labor costs to install them. Fully illustrated with detailed application drawings and photographs.

LEGAL

Construction Law in Contractor's Language
McGraw-Hill Book Co.
Covers such topics as bidding, general contracts, subcontracts, supply contracts, performance, and insurance.

Deskbook of Construction Contract Law
Prentice-Hall, Inc.
Answers to critical legal questions with detailed procedures on how to prepare, present, and resolve contract claims in both federal and non-federal arenas. Also has step-by-step guidelines for avoiding contract disputes before and during construction, model contracts, scheduling clauses, forms, and subcontracts.

Forming Corporations and Partnerships
TAB Books, Inc.
Includes all the necessary forms and step-by-step instructions needed to guide the entrepreneur through forming a corporation, preparing a partnership agreement, or starting a sole proprietorship legally, in every state.

Handbook of Modern Construction Law
Prentice-Hall, Inc.
The complete "answer book" of construction law with standard forms of agreement, sample cases, and

model documents. Shows you exactly how to prevent costly legal trouble in the four places it most often occurs: bid mistakes, change orders, work delays, and payment for work performed.

Labor Law in Contractor's Language
McGraw-Hill Book Co.
Explains labor laws in full detail, precedent and practice.

Lien and Bond Claims in the 50 States
The American Subcontractors Association, Inc.
This manual provides a quick and easy summary of each state's lien and bond laws and the critical deadlines you must follow to guarantee your right to payment.

Subcontractor Bidding and the Law
The American Subcontractors Association, Inc.
Explains in everyday terms the many court decisions that have shaped the law on bidding. This informative manual emphasizes the importance of avoiding bid error by examining all bidding documents, including contracts, bulletins, addenda, and plans and specifications, before submitting a bid.

MASONRY

Architectural and Engineering Concrete Masonry Details
National Concrete Masonry Association
Approximately 150 drawings representing some of the more widely used concrete masonry construction details.

Basement and Foundation Walls
National Concrete Masonry Association
Text, design tables, and suggested construction details are provided to assist in the construction of both nonreinforced and reinforced concrete foundation and basement walls.

Fireplace Design and Construction
John Wiley & Sons, Inc.

Manual of Facts on Concrete Masonry
National Concrete Masonry Association
Compilation of the NCMA series of technical publi-

cations issued monthly covering all facets of concrete masonry.

Masonry & Concrete Construction
Craftsman Book Co.
Every aspect of masonry construction is covered, from laying out the building with a transit to constructing chimneys and fireplaces. Explains footing construction, building foundations, laying out a block wall, reinforced masonry, and much more.

Masons and Builders Library
Audel/Macmillan Publishing Co.
This two-volume set presents practical instruction in bricklaying and masonry. All aspects of materials and methods are outlined in step-by-step procedures illustrated by photographs, diagrams, and tables.

Residential Masonry
Glencoe Publishing Company
A text for teaching masonry skills which focuses on light masonry construction. Each section identifies a major building theme including concrete, concrete block, poured concrete, brick, and stone. Also covers techniques for preparing mortar and laying concrete block and brick.

OFFICE SUPPLIES AND BUSINESS FORMS

The Drawing Board. Offers a wide variety of office supplies and business forms. Free catalog.

Grayarc. Office forms and supplies. Free catalog.

McBee. Selection of one-write bookkeeping systems for construction businesses. Office forms and supplies. Free catalog.

New England Business Systems (NEBS). Offers a wide variety of business forms for contractors, including one-write bookkeeping systems, proposals, work orders, subcontract agreements. Free catalog.

PAINTING AND WALLCOVERING

Paint Contractor's Manual
Craftsman Book Co.
How to start and run a profitable paint contracting company —getting set up and organized to handle volume work, avoiding the mistakes most painters make, getting top production from your crews and the most value from your advertising dollar. Includes man-hour estimates, sample forms, contracts, charts, tables, and examples.

Painter's Handbook
Craftsman Book Co.
Loaded with "how to" information you can use every day to get professional results on any job. Covers every aspect of painting techniques and methods.

Paint Handbook
McGraw-Hill Book Co.
A comprehensive "how to" reference on the various types of paints and coatings and how they are to be applied in a wide variety of areas in industry, commerce, and the home.

Painting and Wallpapering
Ortho
Learn all about wallcoverings and the tools and techniques for applying them.

Painting and Decorating Craftsman's Manual & Textbook
Painting and Decorating Contractors of America
This is the primary reference book for the apprenticeship and training courses in the painting and decorating industry. This 700-page volume is a compilation of eleven smaller books.

Painting and Decorating Encyclopedia
Painting and Decorating Contractors of America
This encyclopedia is a valuable guide and reference work for anyone involved or interested in the industry. Its contents include discussion of materials, tools, surface preparation, and reading blueprints.

Paint Problem Solver
Painting and Decorating Contractors of America
A ninety-page manual addressing major problems faced by painting contractors and the solution to those problems.

Wallcovering Problem Solver
Painting and Decorating Contractors of America
Seventy-five page plus manual fully illustrated to address twenty-five wallcovering problems and their solutions.

PLASTERING

Plastering Skills
American Technical Publishers, Inc.
Describes construction techniques from mixing materials to special surface finishes. Solutions to common plastering problems such as cracking and dryout are given. The how-to chapter on ornamental plastering is of particular interest.

Portland Cement Plaster (Stucco) Manual
Portland Cement Association
Complete guide to plastering and stucco applications. Up-to-date information on materials, bases, mixes, hand and machine application, and curing. Includes glossary of plastering terms.

PLUMBING

Basic Plumbing with Illustrations
Craftsman Book Co.
The journeyman's and apprentice's guide to installing plumbing, piping, and fixtures in residential and light commercial buildings.

Planning and Designing Plumbing Systems
Craftsman Book Co.
Explains with detailed illustrations basic drafting principles for plumbing construction needs. Covers how to use a plot plan, how to convert it into a working drawing, and examples for water supply systems, drainage and venting, pipe, valves, and fixtures.

Plumbers and Pipe Fitters Library
Audel/Macmillan Publishing Co.
This three-volume set contains up-to-date information for master plumbers, journeymen, and apprentices. A detailed text with diagrams, photographs, and charts and tables addresses all aspects of plumbing.

Plumbers Exam Preparation Guide
Craftsman Book Co.
Hundreds of questions and answers to help you pass the apprentice, journeyman, or master plumber's exam. Questions are presented in the style of the actual exam. Answers are given for both standard and uniform plumbing codes.

Plumber's Handbook — Revised
Craftsman Book Co.
Shows what will and will not pass inspection in drainage, vent and waste piping, septic tanks, water supply, fire protection, and gas piping systems.

The Plumbers Handbook
Audel/Macmillan Publishing Co.
A fully illustrated reference guide which covers tools, how to read blueprints, heating systems, water supply, outside sewage lift station, lead work, silver brazing, and much more.

Plumbing Installation and Design
American Technical Publishers, Inc.
Introduces plumbing theory and proven how-to techniques for installing appliances and piping fixtures. Design and installation of a complete residential plumbing system is thoroughly detailed.

Professional Plumbing Techniques — Illustrated and Simplified
TAB Books, Inc.
A field manual or refresher course for professional plumbers. Everything from weights and measures to installation and repair techniques is listed in alphabetical order. An extensive appendix of charts offers measurement tables, pipe sizes, water temperature, friction points, BTU demands, and much more.

Questions and Answers for Plumbers Examinations
Audel/Macmillan Publishing Co.
A practical, fully illustrated study guide for those preparing to take a licensing examination for apprentice, journeyman, or master plumber.

REMODELING

Basic Remodeling Techniques
Ortho
From financing through design and construction, this book shows how to do any residential remodeling job.

The Complete Guide to Remodeling Your Home
Betterway Publications, Inc.
Outlines basic principles and design methods for remodeling residential structures.

Manual of Professional Remodeling
Craftsman Book Co.
Show how to evaluate a job and avoid thirty minute jobs that take all day; what to fix and what to leave alone. Includes detailed information on the various phases of remodel construction.

Remodelers Handbook
Craftsman Book Co.
A complete manual of home improvement contracting — planning the job, estimating the costs, doing the work, running your company, and making the profits. Pages of sample forms, contract documents, clear illustrations, and examples.

Remodelers Handbook: A Manual of Professional Practice for Home Improvement Contractors
National Association of Home Builders
Tells how to evaluate a deteriorated dwelling, plan a job, make structural repairs, remodel kitchens and bathrooms, and create living areas in seldom used space. Also tells how to manage your business, estimate and control costs, sell remodeling, and maintain financial control.

Remodeling Business Basics
National Association of Home Builders
Provides a practical hands-on approach to running a small volume remodeling company. Tells remodelers

how to survive the first five years of business, when the failure rate is high.

ROOFING AND SHEETMETAL

Architectural Sheetmetal Manual
Sheetmetal & Air-conditioning Contractors National Association
The most complete and comprehensive set of recommended practices available for proper design and installation of architectural sheetmetal.

Complete Roofing Handbook
Audel/Macmillan Publishing Co.
This authoritative text, highly detailed drawings, and clear photographs combine to make this the best guide available to all aspects of roofing. Also included are listings of professional and trade associations and of roofing manufacturers.

Handbook of Accepted Roofing Knowledge
National Roofing Contractors Association
A pocket-size booklet that addresses most problems encountered by the roofing contractor and provides recommended practices for roof installation.

Roofer's Handbook
Craftsman Book Co.
The journeyman roofer's complete guide to wood and asphalt shingle application on both new construction and reroofing jobs. Over 250 illustrations and hundreds of trade tips.

NRCA Roofing & Waterproofing Manual
National Roofing Contractors Association
An industry standard of over 600 pages includes information on quality control in built-up roofing, elasto plastic membrane systems design and installation, and single-ply construction details. An extensive reference manual covers decks, vapor retarders, insulation, and much more.

NRCA Construction Details
National Roofing Contractors Association
Seventy-five plates of flushing details for built-up, modified bitumen, and single-ply roof membranes.

Includes details for roof edges, expansion joints, walls, curbs, drains, and penetrations.

Roofing Materials Guide
National Roofing Contractors Association
Contains technical data on roofing and waterproofing products and roof board insulation materials, as well as information on the manufacturers and suppliers of these products. This publication is sold on a subscription basis only, with two issues per year.

Sheet Metal Work
Audel/Macmillan Publishing Co.
An on-the-job guide for workers in the manufacturing and construction industries. All aspects of sheet metal work are explained in detail and illustrated with drawings, photographs, and tables. Covers mathematics for sheet metal work, principles of drafting, layout, bending and folding, soldering and brazing, and much more.

STEEL CONSTRUCTION

Construction of Steel Building Frames
John Wiley & Sons, Inc.

Manual of Steel Construction
American Institute of Steel Construction, Inc.
An essential reference for steel construction. The industry standard for steel members and their design and construction properties.

SURVEYING

Building Layout
Craftsman Book Co.
Shows how to use a transit to locate the building on the lot correctly, plus proper grades with a minimum excavation, finding utility lines and easements, establishing correct elevations, etc.

Construction Surveying and Layout
Craftsman Book Co.
A practical guide to simplified construction surveying. Covers all aspects of surveying in a detail.

WELDING

Arc Welding
American Technical Publishers, Inc.
A well-illustrated guide to the principles and techniques of arc welding in all positions, plus pipe welding and arc cutting.

Practical Welding
Glencoe Publishing Company
Practical, hands-on text helps the student pass welding certification tests and get a job as a welder. Extensive photos and illustrations help describe equipment, welding position, joint design, and welding symbols.

Welders/Fitters Guide
Audel/Macmillan Publishing Co.
Presents step-by-step instruction for those training to become welder/fitters who have some knowledge of welding and the ability to read blueprints.

Welders Guide
Audel/Macmillan Publishing Co.
A manual on the theory, operation, and maintenance of all welding machines. An authoritative text with hundreds of detailed drawings and photographs.

Welding Skills
American Technical Publishers, Inc.
An outstanding textbook guide to welding. Contains the latest techniques, standards, and practices. From theory and safety to practical how-to applications using all modern welding processes and equipment.

Resource Addresses

PUBLISHERS

American Technical Publishers, Inc.
1155 West 175th Street
Homewood, IL 60430
(800) 323-3471

Barnes & Noble Books
126 5th Avenue
New York, NY 10011
(212) 767-8844

Barron's Educational Series, Inc.
250 Wireless
Hauppauge, NY 11788
(718) 762-3430

Betterway Publications, Inc.
P.O. Box 219
Crozet, VA 22932
(804) 823-5661

Craftsman Book Co.
6058 Corte Del Cedro
Carlsbad, CA 92009
(800) 541-0900

Doubleday & Co.
501 Franklin Avenue
Garden City, NY 11530
(516) 294-4000

Enterprise Publishing, Inc.
725 Market Street
Dept. VB72L
Wilmington, DE 19801

Glencoe Publishing Company
15319 Chatsworth Street
Mission Hills, CA 91345
(800) 257-5755

Macmillan Publishing Co. (Audel)
866 Third Avenue
New York, NY 10022
(212) 702-6735

McGrawHill Book Co.
11 West 19th Street
New York, NY 10011
(800) 722-4726

R.S. Means Company, Inc.
100 Construction Plaza
P.O. Box 800
Kingston, MA 02364
(617) 747-1270

The MIT Press
55 Hayword Street
Cambridge, MA 02412
(617) 253-2884

Norseman Publishing Co.
P.O. Box 6617
Lubbock, TX 79493
(800) 843-0213
(800) 795-9875 (Texas only)

Ortho Books
6001 Bollinger Canyon Road, Building T
San Ramon, CA 94583
(415) 842-5537

PAS Publications
3101 E. Eisenhower Parkway, Suite 2
Ann Arbor, MI 48104
(313) 973-6200

Prentice-Hall, Inc.
Route 59 at Brookhill Drive
West Nyack, NY 10995
(201) 767-5937

Safety Meeting Outlines
P.O. Box 294
Park Forest, IL 60466

Self-Counsel Press, Inc.
1303 N. Northgate Way
Seattle, WA 98133
(206) 522-8383

TAB Books, Inc.
Blue Ridge Summit, PA 17214
(800) 233-1128

United States Gypsum Company (USG)
101 South Wacker Drive
Chicago, IL 60606
(312) 321-4221

Van Nostrand Reinhold
115 Fifth Avenue
New York, NY 10003
(212) 254-3232

Frank R. Walker Company
5100 Academy Drive
Lisle, IL 60532
(800) 458-3737

John Wiley & Sons, Inc. Publishers
Occupational Division
605 Third Avenue
New York, NY 10158
(212) 807-0530

ASSOCIATIONS AND ORGANIZATIONS

Air-Conditioning and Refrigeration Institute (ARI)
1501 Wilson Boulevard, 6th Floor
Arlington, VA 22209
(703) 524-8800

The Aluminum Association (AA)
900 19th Street, N.W.
Washington, DC 20006
(202) 862-5100

American Concrete Institute (ACI)
P.O. Box 19150
Detroit, MI 48219
(313) 532-2600

American Institute of Architects (AIA)
1735 New York Avenue
Washington, DC 20006
(202) 626-7475

American Institute of Steel Construction (AISC)
400 North Michigan Avenue
Chicago, IL 60611
(312) 670-5446

American Insurance Association (AIA)
85 John Street
New York, NY 10038
(212) 669-0400

American Iron and Steel Institute (AISI)
1000 6th Street, N.W.
Washington, DC 20036
(202) 452-7100

American National Standards Institute (ANSI)
1430 Broadway
New York, NY 10018
(212) 354-3300

American Society for Testing and Materials (ASTM)
1916 Race Street
Philadelphia, PA 19103
(215) 299-5585

The American Subcontractors Association, Inc. (ASA)
1004 Duke Street
Alexandria, VA 22314
(703) 684-3450

The Association General Contractors of America (AGC)
1957 E Street, N.W.
Washington, DC 20006
(202) 393-2040

Building Official & Code Administrators International (BOCA)
4051 West Flossmoor Road
Country Club Hills, IL 60477
(312) 799-2300

Construction Specifications Institute (CSI)
601 Madison Street
Alexandria, VA 22314
(703) 684-0300

Council of American Building Officials (CABO)
5203 Leesburg Pike, Suite 708
Falls Church, VA 22041
(703) 931-4533

International Association of Plumbing and Mechanical Officials (IAMPO)
5032 Alhambra Avenue
Los Angeles, CA 90032
(213) 223-1471

International Conference of Building Officials (ICBO)
5360 South Workman Mill Road
Whittier, CA 90601
(213) 223-1471

Mason Contractors Association of America (MCAA)
17 West 601 14th Street
Oakbrook Terrace, IL 60181
(312) 620-6767

Mechanical Contractors Association of America (MCAA)
5410 Grosvenor Lane, Suite 120
Bethesda, MD 20814
(301) 897-0770

Metal Lath/Steel Framing Association
600 South Federal, Suite 400
Chicago, IL 60605
(312) 922-6222

National Association of Home Builders (NAHB)
15th and M Streets, N.W.
Washington, DC 20005
(800) 368-5242

National Association of Plumbing, Heating, and Cooling Contractors (NAPHCC)
180 S. Washington St.
P.O. Box 6808
Falls Church, VA 22046
(703) 237-8100

National Concrete Masonry Association (NCMA)
P.O. Box 781
2302 Horse Pen Road
Herndon, VA 22070
(703) 435-4900

National Electrical Contractors Association (NECA)
7315 Wisconsin Avenue
Bethesda, MD 20814
(301) 657-3110

National Fire Protection Association (NFPA)
Batterymarch Park
Quincy, MA 02269-9990
(617) 770-3500

National Insulation Contractors Association (NICA)
1025 Vermont Avenue, N.W., Suite 410
Washington, DC 20005
(202) 783-6277

National Roofing Contractors Association (NRCA)
One O'Hare Center
6250 River Road
Rosemont, IL 60018
(312) 318-6722

Painting and Decorating Contractors of America (PDCA)
7223 Lee Highway
Falls Church, VA 22046
(703) 534-1201

Portland Cement Association (PCA)
5420 Old Orchard Road
Skokie, IL 60077
(312) 966-9599

Prestressed Concrete Institute (PCI)
175 West Jackson Boulevard
Chicago, IL 60604
(312) 786-0300

Research and Education Association (REA)
505 Eighth Avenue
New York, NY 10018
(212) 695-9487

Sheetmetal and Air Conditioning Contractors National Association (SACCNA)
8224 Old Courthouse Road
Vienna, VA 22180
(703) 790-9890

Southern Building Code Congress International, Inc. (SBCCI)
900 Montclair Road
Birmingham, AL 35213-1206
(205) 591-1853

OFFICE SUPPLIES

The Drawing Board
256 Regal Row
P.O. Box 220505
Dallas, TX 75222
(800) 527-9530

Grayarc
Greenwood Industrial Park
P.O. Box 2944
Hartford, CT 06104
(800) 243-5250

McBee
P.O. Box 741
Athens, OH 45701
(800) 526-1272

New England Business Systems (NEBS)
500 Main Street
Groton, MA 01471
(800) 225-6380

GOVERNMENT AGENCIES

Bureau of Census
Customer Service
Washington, DC 20238

Federal Trade Commission
Division of Legal & Public Records
Washington, DC 20580

Internal Revenue Service
Washington, DC 20224

Library of Congress
National Referral Center
Washington, DC 20559

National Bureau of Standards
Technical Building B167
Standards Development Services Section
Washington, DC 20234

SBA Management Assistant Publications
P.O. Box 15434
Fort Worth, TX 76119

SCORE — Service Corps of Retired Executives
1441 L Street, N.W., Room 100
Washington, DC 20416

Superintendent of Documents
U.S. Government Printing Office
Washington, DC 20402

U.S. Dept. of Labor
200 Constitution Avenue, N.W.
Washington, DC 20210

U.S. Small Business Administration (SBA)
1441 L Street, N.W.
Washington, DC 20416

BIBLIOGRAPHY

Brabec, Barbara. *Homemade Money*. Crozet, VA: Betterway Publications, Inc.

Cornish, Clive G., CPA. *Basic Accounting for the Small Business*. Seattle: Self-Counsel Press, Inc.

Estes, Ralph. *Dictionary of Accounting*. Cambridge, MA: The MIT Press.

Fields, Louis W. *Bookkeeping Made Easy*. Garden City, NY: Doubleday & Co., Inc.

Gerber, Michael E. *The E Myth*. Cambridge, MA: Ballinger Publishing Co., a subsidiary of Harper & Row Publishers, Inc.

Klotzburger, Katherine M., Ph.D. *How to Qualify for the Home Office Deduction*. Crozet, VA: Betterway Publications, Inc.

How to Subcontract and Supervise Hired Personnel. Washington, DC: National Association of Home Builders.

Practical Accounting and Cost Keeping for Contractors. Lisle, IL: Frank R. Walker Company, Inc.

INVENTORY

Inventory Date _____ Taken By _____ Sheet _____ of _____

M#	DESCRIPTION	QUAN	UNIT	UNIT PRICE	EXTENT

FIGURE 1

PURCHASING WORKSHEET

Page ____ of ___

Job _____   Address _____

Scheduled Start Date _____

QTY.	UNIT	DESCRIPTION	EST. PRICE	UNIT PRICE	TOTAL COSTS	DATE PURCH.	DUE ON JOB	PURCH. FROM

FIGURE 2

ALIGN TYPE DIRECTLY ON THIS LINE _____

PRODUCT 924 /NEBS/ Inc., Groton, Mass 01471 To Order PHONE TOLL FREE 1 + 800-225-6380

PURCHASE ORDER

Show this Purchase Order Number
on all correspondence, invoices,
shipping papers and packages.

TO

DATE	REQUISITION NO.
SHIP TO	

REQUISITIONED BY	WHEN SHIP	SHIP VIA	F.O.B. POINT	TERMS	
QTY. ORDERED	QTY. RECEIVED	STOCK NO. / DESCRIPTION		UNIT PRICE	TOTAL

1. Please send _____ copies of your invoice.
2. Order is to be entered in accordance with prices, delivery and specifications shown above.
3. Notify us immediately if you are unable to ship as specified.

AUTHORIZED BY _____

ORIGINAL

PROJECT SCHEDULING

PROJECT	ACTIVITY	WORKERS	EST. START	EST. COMPT.	NO. OF DAYS	REMARKS

FIGURE 4

Approximate Areas of Model Code Influence

BID WORKSHEET

PROJECT	CLIENT		DATE
LOCATION	ADDRESS		ESTIMATED BY
ARCH/ENGR			EXTENDED BY
PROJECTED DAYS TO COMP.	PHONE		DATE BID DUE

DESCRIPTION	QTY.	UNIT	UNIT	MATRL	LABOR	EQUIP	TOTAL
TOTALS							

RECAP

TOTAL MATERIAL & LABOR COST	OVERHEAD (%)
MISC. OVERRUNS	PROFIT (%)
TOTAL PROJECT COST	TOTAL BID PRICE

FIGURE 6

JOB EXPENSE REPORT

(OWNER):	PROJECT:	PERIOD FROM:
ATTENTION:	LOCATION:	TO:
		CONTRACT DATE:

DATE	DESCRIPTION	LABOR HOURS	LABOR UNIT PRICE	MATRL	LABOR	EQUIP	MISC.	PAYMENT
	SUB-TOTALS							

RECAP

Expenses Prior to This Report...$_____

Current Expenses...$_____

Total Expenses to Date..$_____

Add _____ %...$_____

Cost Plus Sum to Date...$_____

Total Estimated Costs...........................$_____ Est. Percentage of Comp. _____%

Less Previous Payments...$_____

Current Payment Due..$_____

FIGURE 7

SUPPLIER BID SHEET

Sheet ____of ___

PROJECT _____ LOCATION _____ BID DATE_____

DATE	SUPPLIER/SUB	USED	DESCRIPTION	INC. LABOR	INC MATL.	INC. TAX & FT	QUOTE

FIGURE 8

Telephone Bid

JOB NAME:	TIME:	AM	PM
	DATE:		

FIRM QUOTING

BID GIVEN BY

ADDRESS

PHONE

CLASS OF WORK

JOB LOCATION

ESTIMATE NO

BID TAKEN BY

ACKNOWLEDGEMENT OF ADDENDA

WORK TO BE PERFORMED	AMOUNT OF BID

BOND INCLUDED

SALES TAX INCLUDED

DELIVERED TO JOB SITE

MISC.

PRODUCT 238 NEBS Inc., Groton. Mass. 01471. To Order PHONE TOLL FREE 1 + 800-225-6380

Bid Comparison

FIRM QUOTING	AMOUNT OF BID		
	JOB NAME		
	CLASS OF WORK	ESTIMATE NO	
		DATE	
		REMARKS	

COMMENTS

XYZ MECHANICAL CONTRACTORS
1234 MAIN STREET
SPRINGTOWN, ARKANSAS 34567

March 10, 1988

Mark S. Wolfe
3456 South Street
Marshall, Arkansas 34576

Dear Mr. Wolfe:

It was really a pleasure meeting you Tuesday morning. I hope that this is only the beginning of a long and mutually rewarding relationship.

Thanks for allowing us the opportunity to bid the mechanical portion in the Freestone Plant Facility. I am enclosing a copy of our proposal on this project. If you have any questions, I will be glad to go over the proposal with you.

I am also enclosing a list of some of the contractors and jobs that we have done work for within the last two years. Please feel free to inquire about our workmanship and service with any of these companies.

Again, let me state my desire to work with you and your firm on this and other projects in the near future.

Sincerely,

Paul D. Scott
President

PDS/ab

Enclosures

ESTIMATE

CASEY INTERIORS
5201 S. Main
Townsville, USA
(123) 444-0000

February 1, 1988

ESTIMATE FOR:

Johnson Builders
421 Front Street
Townsville, USA

Re: Preliminary estimate for Towers Restaurant located at 3009 High Drive,
    Townsville, USA.

   This is a preliminary estimate for the interior finish work in the
Towers Restaurant remodel. This estimate is based on preliminary plans by
Weir Architects and according to information given during the January 21,
1988 meeting with Samuel Weir, Charles Johnson, and myself.

   The price includes drywall, wallpaper, painting as noted, interior trim
work, and suspended acoustical ceiling in the office space.

   The estimated price is $15,768.00. This price is subject to change based
on the results of the final design by Weir Architects. This is NOT a fixed
price. This is only an estimate.

Sam Casey
Casey Interiors

𝔓𝔯𝔬𝔭𝔬𝔰𝔞𝔩

Page No.　　　 of 　　　 Pages

YOUR FIRM NAME HERE
123 Main Street
YOUR TOWN, STATE AND ZIP
Phone 123-4567

PROPOSAL SUBMITTED TO		PHONE	DATE
STREET		JOB NAME	
CITY, STATE AND ZIP CODE		JOB LOCATION	
ARCHITECT	DATE OF PLANS		JOB PHONE

We hereby submit specifications and estimates for:

𝔚𝔢 𝔓𝔯𝔬𝔭𝔬𝔰𝔢 hereby to furnish material and labor — complete in accordance with above specifications, for the sum of:

_____ dollars ($ _____).

Payment to be made as follows:

All material is guaranteed to be as specified. All work to be completed in a workmanlike manner according to standard practices. Any alteration or deviation from above specifications involving extra costs will be executed only upon written orders, and will become an extra charge over and above the estimate. All agreements contingent upon strikes, accidents or delays beyond our control. Owner to carry fire, tornado and other necessary insurance. Our workers are fully covered by Workmen's Compensation Insurance.

Authorized
Signature _____

Note: This proposal may be
withdrawn by us if not accepted within_____ days.

𝔄𝔠𝔠𝔢𝔭𝔱𝔞𝔫𝔠𝔢 𝔬𝔣 𝔓𝔯𝔬𝔭𝔬𝔰𝔞𝔩 —The above prices, specifications and conditions are satisfactory and are hereby accepted. You are authorized to do the work as specified. Payment will be made as outlined above.

Signature _____

Date of Acceptance: _____

Signature _____

CHANGE ORDER

(OWNER):	PROJECT:	CHANGE ORDER NO:
ATTENTION:	LOCATION:	
		DATE:

This change order is hereby incorporated into the original contract and is to be attached thereto. All other items and conditions of the original contract and any prior change orders not modified below remain the same. Changes are as follows:

ITEM #	DESCRIPTION OF CHANGE	ADDITIONS	DEDUCTIONS
	TOTALS		
	NET CHANGE BY THIS CHANGE ORDER		

Original Contract Sum ....$_____
Net Change by Previous Change Orders ....$_____
Net Change by This Change Order ....$_____
Revised Contract Sum..$_____
This Change Order Increases the date of substantial completion of this job by _____ days.

Accepted By: _____
              Contractor

              _____
              Owner

FIGURE 13

FINANCIAL STATEMENT

Date _____

ASSETS

1. CASH ON HAND   5,162.00

2. SECURITIES
300 shares of Acme stock at $2.00/share   600.00

3. EQUIPMENT
'81 Dodge truck   3,250.00
16 ft trailer   625.00
8 hp generator   550.00
Air compressor & guns   825.00
Total equipment value   5,250.00

4. REAL ESTATE
Personal home appraised value   62,500.00
Lot purchased in Smithville addition   5,750.00
Total real estate value   68,250.00

5. CURRENT CONTRACTS
Accounts receivable   35,100.00
TOTAL ASSETS $114,362.00

LIABILITIES

1. EQUIPMENT
Note payoff on Dodge truck
     (insured for $3,500.00)   1,720.00

2. REAL ESTATE
Mortgage payoff on personal home   41,750.00
Note payoff on lot in Smithville addition   2,905.00
Total real estate liability   44,655.00

3. CURRENT CONTRACTS
Accounts payable   22,378.00
TOTAL LIABILITIES $68,753.00

NET WORTH

NET WORTH (Assets less Liabilities) $45,609.00

APPLICATION FOR DRAW

TO:	ATTENTION:
PROJECT	DRAW#

ITEM #	DESCRIPTION OF WORK	COMPLETED	TOTAL DRAWN TO DATE

Distribution to:
❑ OWNER
❑ BUILDER
❑ BANK

Application is made for draw, as shown below, in connection with this contract.

TOTAL COMPLETED
   & STORED
      TO DATE........................$_____
LESS PREVIOUS
      DRAWS.........................$_____

CURRENT DRAW...............$_____

BALANCE LEFT TO
DRAW ON THIS
PROJECT...........................$_____

The undersigned builder certifies that to the best of his knowledge, information and belief the work covered by this Application for Draw has been completed in accordance with the Contract Documents, that all amounts have been paid by him for work for which prevous draws were issued, and that current payment shown herein is now due.

CONTRACTOR_____

BY:_____ DATE:_____

BUSINESS RESUME
for
BROWN CONSTRUCTION COMPANY
1056 Commerce Street
Mainville, Arkansas 75032

OWNER: Robert E. Brown
1056 Commerce Street
Mainville, Arkansas 75032
(123) 333-0000

OWNER'S EXPERIENCE:

1984 - Present	Owner: Brown Construction Company. Responsible for bidding, purchasing, job supervision, sales, and personnel.
1978 - 1984	Job Superintendent: Clyde S. Spenser Contractors, Little Rock, Arkansas. Responsible for managing large construction projects for carpentry and interior finish-out. Also involved in bidding.
1971 - 1978	Job foreman: Handle Framing Contractors, Inc., Little Rock, Arkansas. In charge of residential framing, light commercial metal stud framing, worked with city building inspectors, and purchased materials for the jobs.
1964 - 1971	Carpenter Journeyman: Dalworth & Co., Jonesville, Arkansas. Worked up from carpenter's helper to journeyman in charge of framing crew.
BUSINESS HISTORY:	Brown Construction Company started in June 1984 with two helpers and myself doing the work. We have grown to 15 full-time and 2 part-time employees doing interior finish-out and commercial remodel work as well as commercial framing.

PROJECTS BY BROWN CONSTRUCTION:

Billie-Jean's Restaurant, Hwy 74, Mainville, AR. Remodel of 3,500 sq ft dining area. AMX Contractors.

Davis Office Complex, Main Street, Mainville, AR. Framing and drywall for new construction of 8,000 sq ft office complex. Bill Davis.

Robert W. Fox, Hillside Dr., Mainville, AR. Addition of 500 sq ft to Fox residence. Robert Fox.

Metro-Medical Center, 4th Ave., Little Rock, AR. 10,000 sq ft finish-out which included interior wall framing, drywall, and suspended acoustical ceiling. Sherman Pots Construction, Inc.

TRADE REFERENCES:

Maybank Lumber Co., P.O. Box 413, Mainville, AR (123) 234-9876
B & F Building Supplies, P.O. Box 3214, Mainville, AR (123) 222-0000
White Drywall Supplies, 140 Main Street, Mainville, AR (123) 222-1111

APPLICATION FOR EMPLOYMENT

NOTE: If you require more space than provided, please attach separate sheet(s).

PERSONAL

NAME		TODAY'S DATE

STREET	CITY	REFERRED BY:

STATE	ZIP	SOCIAL SECURITY NUMBER	APPLYING FOR:

☐ FULL TIME   ☐ PART TIME   ☐ TEMPORARY

HOME PHONE	BEST TIME TO CALL	BUSINESS PHONE	BEST TIME TO CALL

EDUCATION

	NAME AND LOCATION	FROM	TO	CURRICULUM		DATE GRADUATED
HIGH SCHOOL						
COLLEGE				MAJOR	DEGREE	
OTHER						

SPECIAL SKILLS OR TRAINING (That May Qualify You For Work With Our Company)

EMPLOYMENT (Start With Most Recent)

FROM	TO	EMPLOYER	PHONE ( )	CITY, STATE
JOB TITLE		DUTIES		
SUPERVISOR'S NAME				
STARTING SALARY / WAGES				
FINAL SALARY / WAGES		REASON FOR LEAVING		
FROM	TO	EMPLOYER	PHONE ( )	CITY, STATE
JOB TITLE		DUTIES		
SUPERVISOR'S NAME				
STARTING SALARY / WAGES				
FINAL SALARY / WAGES		REASON FOR LEAVING		
FROM	TO	EMPLOYER	PHONE ( )	CITY, STATE
JOB TITLE		DUTIES		
SUPERVISOR'S NAME				
STARTING SALARY / WAGES				
FINAL SALARY / WAGES		REASON FOR LEAVING		
FROM	TO	EMPLOYER	PHONE ( )	CITY, STATE
JOB TITLE		DUTIES		
SUPERVISOR'S NAME				
STARTING SALARY / WAGES				
FINAL SALARY / WAGES		REASON FOR LEAVING		

POSITION(S) DESIRED

HOURS / DAYS AVAILABLE

U.S. MILITARY RECORD

BRANCH OF SERVICE	FROM	TO	DUTIES	DISCHARGE DATE

REFERENCES

NAME	ADDRESS	YEARS KNOWN

APPLICANT'S STATEMENT

I certify that statements made by me on this form are true and correct. I understand that if employed, any false statement on this application can be considered cause for dismissal. I authorize investigation of all statements contained in this application for employment as may be necessary in arriving at an employment decision.

Signature _____   Date _____

DO NOT WRITE BELOW THIS LINE

— —

PERSONNEL ACTION

REMARKS:

PRODUCT 505-1 /NEBS/ Inc., Groton, Mass. 01471. To Order PHONE TOLL FREE 1 + 800-225-6380 (Mass. residents 1 + 800-252-9226)

```
                  XYZ MECHANICAL CONTRACTORS
                       1234 MAIN STREET
                  SPRINGTOWN, ARKANSAS 34567




May 25, 1988


Mark S. Wolfe
3456 South Street
Marshall, Arkansas 34576

Dear Mr. Wolfe:

    Thank you for the opportunity to work for your organization on the pro-
ject at the Freestone Plant Facility. It is always a pleasure to do busi-
ness with a firm that is as organized and courteous to the subcontractors
on the job as your personnel was to us.

    We hope that we have pleased you with our phase of the job and that you
will consider us again when you have need for a mechanical contractor. If
there is any way that we can be of assistance to you, please do not hesi-
tate to give us a call.


Sincerely,



Paul D. Scott
President


PDS/ab
```

DeVALT CONCRETE CONTRACTORS
490 FIELDER
MABANK, OHIO 68743

January 23, 1988

Bob T. Fender
James Construction Company
9990 Land Avenue
Mabank, Ohio 68744

Dear Bob,

I have recently heard of your move to James Construction Company, and want to be among the first to congratulate you on this new business opportunity. I'm sure that you will be an asset to their firm, just as you were at Grayson-Weir.

I feel that we have enjoyed a good working relationship with you through the years and hope that you will allow us to continue this relationship in your new position. In this fast-paced industry, it is easy to lose contact with those whom you are not in touch with on a regular basis, and I sincerely hope that this will not be the case with us.

Wishing you the best in your new position, and hoping to be hearing from you soon, I am

Yours sincerely,

Joe B. Wheeler
President

JBW/sm

CHECK REGISTER

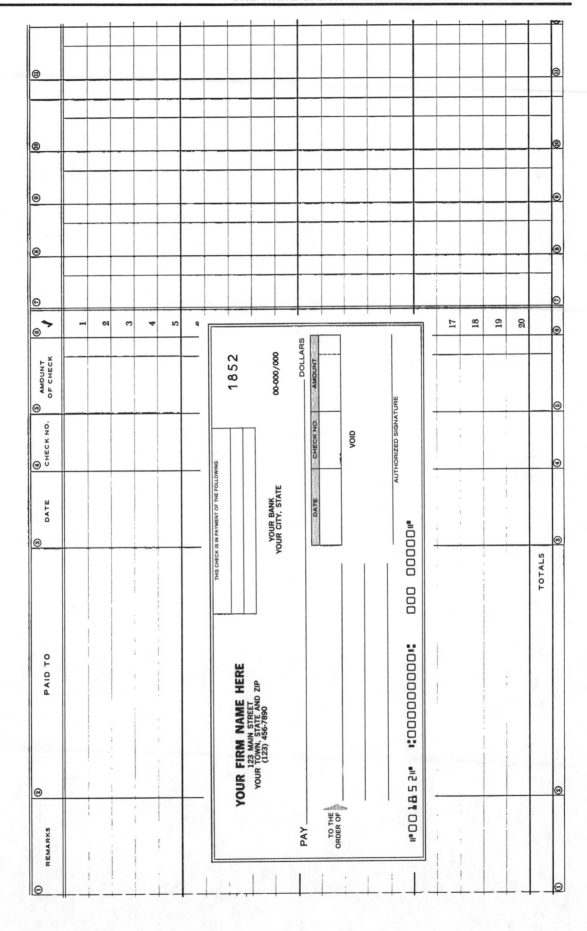

BANK BALANCE

BANK NAME

DATE

FOR DEPOSITS FOR

SCRIPTION CK NOS.

$ TOTAL OF CHECKS WRITTEN

AMOUNT DEPOSITED

BANK BALANCE

BANK BALANCE FORWARD

CHECK REGISTER FOR MONTH OF ____ 19

PAGE NO.

CHECK NOS. FROM TO

BANK ACCOUNT RECONCILIATION

Balance Shown on Bank Statement

List Below and Subtract all Outstanding Checks

DATE OR CHECK NO. | CHECK AMOUNT | DATE OR CHECK NO. | CHECK AMOUNT

TOTAL LEFT-HAND COLUMN —

SUB-TOTAL

ADD DEPOSITS NOT CREDITED +

ADD BANK SERVICE CHARGES +

OUR MONTH-END BANK BALANCE =

If Balance does not prove, show adjustments on these lines.

OUR MONTH-END BANK BALANCE =

NOTES OR REMARKS

EARNINGS
RECORD OF:
ADDRESS

NO. OF EXEMPT.		MARRIED		SOC. SEC. NO.	
PHONE		SINGLE			
		DATE EMPLOYED		IN EMERGENCY, NOTIFY:	

NEBS Inc., Groton Mass. 01471.
PRODUCT 460

FIXED PAYROLL

PERIOD ENDED	HRS. WORKED		EARNINGS				DEDUCTIONS						NET PAY	NAME, INITIALS OR CHECK NO.
	REG.	O.T.	REGULAR	OVERTIME	OTHER	TOTAL	F.I.C.A.	FEDERAL WITH. TAX	STATE TAX	A	B	C		
1														
2														
3														
4														
5														
6														
7														
8														
9														
10														
11														
12														
13														
14														
15														
TOTAL														
YR. TO DATE														
1														
2														
3														
4														
5														
6														
7														
8														
9														
10														
11														
12														
13														
14														
15														
TOTAL														
YR. TO DATE														

PAY STATEMENT

PERIOD ENDED — HRS. WORKED (REG., O.T.) — EARNINGS (REGULAR, OVERTIME, OTHER, TOTAL) — DEDUCTIONS (F.I.C.A., FEDERAL WITH. TAX, STATE TAX, A, B, C) — NET PAY — NAME OR INITIALS

THIS IS A STATEMENT OF YOUR EARNINGS AND DEDUCTIONS AS REPORTED TO THE FEDERAL AND STATE GOVERNMENTS. RETAIN IT PERMANENTLY FOR YOUR TAX RECORDS.

TIME CARD

WEEK ENDING _____ 19___

EMPLOYEE'S NAME

ADDRESS

SOCIAL SECURITY NO.

SIGNATURE

NUMBER OF DEPENDENTS

EARNINGS	AMOUNT
REGULAR HOURS @	
OVERTIME HOURS @	
GROSS EARNINGS $	
DEDUCTIONS	
SOCIAL SECURITY	
FED. TAX	
STATE TAX	
INSURANCE	
TOTAL DEDUCTIONS $	
NET EARNINGS $	

CHECK NO. _____ DATE _____ 19___

	A.M.		P.M.		HOURS WK'D.
	IN	OUT	IN	OUT	
MON.					
TUE.					
WED.					
THU.					
FRI.					
SAT.					
SUN.					
TOTAL					

JOB NAME OR NUMBER — REG. HOURS — OVER-TIME

ABOVE SECTION MAY BE USED TO LIST MATERIALS, MISC. EXPENSES, SUNDAY WORK, ETC.

MONDAY TOTAL ▶

TUESDAY TOTAL ▶   DATE

THURSDAY TOTAL ▶   DATE

FRIDAY TOTAL ▶

SATURDAY TOTAL ▶   DATE

PRODUCT 225   *NEBS* Inc. Groton, Mass. 01471. To Order PHONE TOLL FREE 1 + 800-225-6380

JOB COST RECORD

PROJECT _____

DATE	DESCRIPTION	MATERIAL	LABOR	EQUIPMENT	MISC.
	TOTALS				

PERCENTAGE OF COMPLETION ...._____%
TOTAL ESTIMATED COST (including changes) ....$_____
ACTUAL EXPENSES TO DATE..$_____
VARIANCE ....$_____

FIGURE 24

INVOICE

TO _____

DATE _____ JOB NO. _____

JOB NAME _____

JOB LOCATION _____

TERMS

	DESCRIPTION	PRICE	AMOUNT

ORIGINAL

Thank You

STATEMENT

YOUR FIRM NAME HERE
123 Main Street
YOUR TOWN, STATE AND ZIP

Phone 123-4567

DATE _____

TERMS:

PLEASE DETACH AND RETURN WITH YOUR REMITTANCE $ _____

DATE	INVOICE NUMBER / DESCRIPTION	CHARGES	CREDITS	BALANCE
	BALANCE FORWARD ⇨			

YOUR FIRM NAME HERE

Thank You   ⇧ PAY LAST AMOUNT IN THIS COLUMN

ABC PAINTING CONTRACTORS
555 AVENUE J
FORT GREENE, WYOMING 76590

July 20, 1988

Jack O. Rose
Hyatt & Greer, Inc.
4690 Miller Road
Fort Greene, Wyoming 76590

Dear Mr. Rose:

This is just a reminder that your account is past due. We are aware that circumstances sometimes arise that cause a company to fail to remit their payment on time.

The amount in question is $1,543.00. We ask that you please remit this amount today.

If you have already mailed your payment, please disregard this letter and accept our thanks.

Sincerely,

Jim E. Gray
President

JEG/mm

ABC PAINTING CONTRACTORS
555 AVENUE J
FORT GREENE, WYOMING 76590

July 1, 1988

Jack O. Rose
Hyatt & Greer, Inc.
4690 Miller Road
Fort Greene, Wyoming 76590

Dear Mr. Rose:

   We regret that we have to send you this _____ notification of your
outstanding balance of $1,543.00, but as of this date, we have not heard
from you.

   Please send your payment today, so that we can clear up this account and
will not be forced to take any additional steps toward collection.

   If payment has been made, please ignore this letter.

Sincerely,

Jim E. Gray
President

JEG/mm

```
                    ABC PAINTING CONTRACTORS
                          555 AVENUE J
                    FORT GREENE, WYOMING 76590
```

July 20, 1988

Jack O. Rose
Hyatt & Greer, Inc.
4690 Miller Road
Fort Greene, Wyoming 76590

Dear Mr. Rose:

   Since you have failed to respond to our previous requests for payment of
your past due account of $1,543.00, we must now insist that you send us the
check for this amount.

   If we have not heard from you in five (5) business days, we will be
turning your account over to a collection agency. We hope that this step
will not be necessary, because we're sure that neither of us wants to see
your credit standing jeopardized.

   If you have made this payment, please ignore this letter. If your payment has not been made, we are urging you to do so immediately.

Sincerely,

Jim E. Gray
President

JEG/mm

ACCOUNTS RECEIVABLE JOURNAL

Page ____ of ____

INVOICE DATE	DESCRIPTION	AMOUNT	DATE DUE	DATE REC'D
	TOTALS			

FIGURE 30

CASH RECEIPTS JOURNAL

Page ____ of ____

DATE REC'D.	DESCRIPTION	AMOUNT
		TOTALS

FIGURE 31

ACCOUNTS PAYABLE JOURNAL

Page ____ of ____

DATE REC'D	ACCOUNT	PROJECT	AMOUNT	DISCOUNT DUE DATE	DATE PAID

FIGURE 32

INDEX

A

Accountant, 20, 98, 114, 129
Accounting, 97-114
    terms, 98-103
Accounts paid, 98
Accounts payable, 98
Accounts payable journal, 109
Accounts payable ledger, 108
Accounts receivable, 98, 106
Accounts receivable journal, 107
Accounts receivable list, 106
Advertising, 26
American Insurance Association, 20
Amortization, 98
Answering machines, 34
Appeal letter, 108
Application for draw, 71
Application for employment, 76
Appreciation, 98
Architects, 40
Asset, 71, 98-9, 127
Assumed Name Certificate, 16
Attorney, 20, 68, 115-6, 129
Audit, 99
Automobile insurance, 118-9
    collision, 118-9
    liability coverage of, 118

B

Balance sheet, 99
Bank note, 70
Banker, as source for job referrals, 26
Banking terms, 70-1
Bankruptcy, 99, 127
Bid bond, 119
Bid worksheet, 51, 52, 59, 72
Bidding, 52-60
    client considerations in, 56
    cost plus, 53-5
    fixed price, 51-2
    guidelines for, 57-8
    unit price, 55-6
        advantages of, 55-6
        disadvantages of, 56
Bids, submitting, 59-60
Bids, telephone, 59

Billing, 106-8
Bills, inability to pay, 91-2
Bills, paying, 108
Blueprint reading, 39-40
Bond, bid, see bid bond
Bond, payment, see payment bond
Bond, performance, see performance bond
Bonds, 66, 119
Bookkeeping services, 34
Budget variance, 99
Building codes, 42, 45-8
Building Officials and Code Administrators International, 46
Bureau of the Census, 27
Business, 15, 126-7
    beginning on part-time basis, 15
    quitting the, 126-7
Business functions, 22
Business resume, 72
Business system, 21-2

C

Capital, 99
Capital asset, 99
Cash disbursements journal, 99, 103-5
Cash flow, 99, 106
Cash receipts journal, 99, 100, 108
Certificate of Insurance, 118
Certification, 20
Certified Public Accountant, 114
Change order, 65-6, 67
Character, 72
Chart of accounts, 99
Check register, 99, 103-105
Client referrals, 26
Clients, 83-9
    contacting, 16
    employees of, 88-9
    entertaining, 85-6
    problem situations with, 86-7
    quoting prices to, 84
    selling yourself to, 83-4
    super rich, 88
    terminating relationship with, 88
Collateral, 70

Commission for obtaining jobs, 16
Commitment to obligations, 86
Company name, 16, 115
Computers, 36-7
Con artists, 87
Contract amount, 65
Contracts, 61-8, 116
    negotiating, 68
    terms of payment, 65
    verbal, 116
    verbal, 61
    written, 61, 62
Corporation, 17, 20
    advantages of, 17, 20
    board of directors, 20
    disadvantages of, 17, 20
    forming, 20
Correspondence, informal, 33-4
Costs, controlling, 29
Council of American Building Officials, 48
Cover letter, 59-60
Credit, 71, 99

D

Daily report, 45
Debit, 71, 99
Delay, 66
Demand letter, 108
Depreciation, 101
Diary, 33
Direct cost, 101
Discount, 101, 108
Diversification, 125
Division of responsibilities, 19-20
Dodge Reports, 26-7
Draw, 71
Dress and appearance, 86

E

Employee earnings record, 104
Employees, 75-81
    assessing talents of, 77
    dismissal of, 79-80
    family and friends as, 77-8
    files of, 79-80
    former, 80

Employees, cont.
   hiring, 75-6
   independence of, 78
Employer extra FICA, 111
Employer Identification Number, 111
Employer's Tax Guide (Circular E), 110, 111, 112
Employment Eligibility Verification Form (Form I9), 80, 81
Engineers, 40
   hiring, 41
Equipment, 30
   leasing, 30-1
     advantages of, 30-1
     disadvantages of, 31
   purchasing, 30
Equity, 101
Estimates, 62
Estimating and bidding, 51-60
Expense report, 55

F

Family, 123
Federal Bankruptcy Act, 99
Federal Identification Number, 16, 17
Federal Information Center, 27
Federal Unemployment Tax, see FUTA
FICA, 110
Files, setting up, 32
Financial statement, 71, 72, 98, 99, 101, 102, 119
Financing, 69-72
Form 8109, 111
Form 940, 111
Form 941, 111
Form SS-4, 16
FUTA, 111
Future planning, 125-7

G

General ledger, 101
Government agencies, 27-8
Gross profit, 101
Gross sales, 101

H

Hold harmless agreement, 66
Home environment, 123
Honesty in business, 89

I

Illegal immigrant workers, 80-1
Immigration Reform and Control Act of 1986, 80, 81
   penalties for noncompliance with, 80-1
Income, 101
Income statement, 101
Indemnification, 66
Indirect cost, 101
Inspectors, 48
Insurance, 66, 117-9
   premiums, 117
   automobile, see automobile insurance,
   duplication of coverage, 119
   job classification and, 117
   liability, see liability insurance,
   payroll audit for, 117
Interest expense, 101
Internal Revenue Service, 16-7, 110-11
International Conference of Building Officials, 46
Inventory, 101
Invoice, 44, 55, 71, 102, 106
Invoices, overdue, 106, 108
Itemized deductions, 102

J

Job cost, 104-6
Job cost accounting, 102
Job cost payroll, 105
Job cost record, 54, 102, 105
Job expense report, 105
Job site, 26, 44-5, 85
   children or pets on, 85
   drinking on, 85
   friends on, 85
   management, 44-5
   parking on, 85
   visits, 26
Journal, 102

L

Labor laws, 76-7
Lawsuits, 115
Legal advice, 115-6
Legal forms of business, 17-20
Letter of appreciation, 84
Letter of congratulations, 89
Liabilities, 71
Liability, 102
Liability insurance, 118
Library of Congress National Referral Center, 27
Licenses, 20
Lien, 70, 108, 115
   mechanic's, 115
Liquid assets, 102
Loans, sources for, 69-70

M

Market expansion, 126
Marketable securities, 102
Material storage, 31
Material suppliers, 91-3
   as information source, 92
   as source for job referrals, 26
   bids from, 58-9
   service provided by, 43
Materials, checking shipment of, 44
Materials, 41-3
   costs, 43
   delivery costs, 43
   inventory of, 42-3
   knowledge of, 41-2
   purchasing, 42-4
   quality, 43
   quantity, 43
   rate of use, 43
   storing, 43
McGraw-Hill Informations Systems Company, 26, 42
Mechanic's lien, see lien, mechanic's
Miscellaneous expenses, 103
Mobile telephones, 36
   leasing, 36
   purchasing, 36
Model Energy Code, 48

N

Net worth, 102, 127
Note payable, 102
Note receivable, 102
National Association of Small Business Investment Companies, 70
National Codes, 46
National Electric Code, 47
National Fire Protection Association, 47
Notary, 68

O

Occupational Safety and Health Act (OSHA) of 1970, 76-7
Office organization, 32-4
Office supplies, 33
One and Two Family Dwelling Code, 47
One-write system, 102, 103
Operational procedures, 22
Operations manual, 22-3
Organizational structure, 21-4
Overhead, 29, 53, 126

P

Pagers, 34-6
    alphanumeric display, 35
    digital display, 35
    leasing, 35
    purchasing, 35
    tone, 35
    used with answering services, 35
    voice, 35
Partnership, 17, 18-9
    advantages of, 17, 18-9
    contract for, 18
    disadvantages of, 17, 18-9
    general, 17, 18-9
    limited, 17, 18-9
Pay statement, 104
Payment bond, 119
Payment schedule, 64
Payroll journal, 102
Payroll procedures, 104
Peers, advice from, 127, 129
Pegboard system, see one-write system

Performance bond, 119
Personal property, 102
Planning, 15
Plans, changes in, 40-1
Preparing to start business, 15
Principal, 102
Priority lists, 24-5
Privacy, need for, 33
Professional courtesy, 15
Profit, 53
Profit and loss statement, 102
Proposal, 61-8
    acceptance of, 65
Proposal form, 62, 63
Public contracts, 21
Purchase order, 44
Purchase records, 44
Purchasing worksheet, 44, 59

Q

Quality of work, 23, 39
Quotes, written, 59

R

Real property, 102
Reconciliation, 102
Record keeping, 97-114
References, 76
Reminder, 108
Retainage, 58, 65
Revenue, 102

S

Scope of work, 65
SCORE, 28
Secretarial services, 34
Secretaries, 34
Service Corps of Retired Executives, see SCORE
Side jobs, 15, 16
Site conditions, 58
Small Business Administration, 27
    Answer Desk, 27
    disaster lending program, 27
    loans, 70
    publications, 27
Social Security, 110

Social Security Act of 1935, as amended, 77
Sole proprietorship, 17-8
    advantages of, 17, 18
    disadvantages of, 17, 18
Southern Building Code Congress International, Inc., 46
Standard Codes, 46
Statement, 106
Subchapter S corporation, 17, 20
    advantages of, 17, 20
    disadvantages of, 17, 20
Subcontractor/supplier bid sheet, 58-9
Subcontractors, 91-3
    as information source, 92
    bids from, 58-9
    hiring, 92-3
Summary of State Regulations and Laws Affecting Contractors, 20
Surety bonds, 21
Sweets Catalogs, 42

T

Tax deductions, 112-14
Taxes, 98, 110-14
    income, 21
    local, 21
    payroll, 110-11
        how to pay, 111
        when to pay, 111
    property, 112
    self-employment, 112
    state, 112
    withholding, 110
Telephone, business, 33
Telephone bid sheet, 59
The Fair Labor Standards Act of 1938, as amended, 76
Time card, 55, 105
Time management, 24-6
Time management planning systems, 25
Trial balance, 103

U

U.S. Department of Labor, 17, 77
    Wage & Hour Division, 76

U.S. Immigration and Naturaliza-
    tion Service, 81
Uniform Codes, 46
Uniform Partnership Act, 17
Uninsured motorist policy, 119

V

Vehicles, 29-30

W

W-2, 112
W-4, 110
Warehouses, 31-2

Warranty, 64, 68, 93
Warranty work, 39
Work schedule, 45
Work, finding, 26-7
Workload, 23, 75, 79
Workmen's compensation, 117-8
    subcontractors and, 117-8